Studies in Systems, Decision and Control

Volume 600

The series "Studies in Systems, Decision and Control" (SSDC) covers both new developments and advances, as well as the state of the art, in the various areas of broadly perceived systems, decision making and control–quickly, up to date and with a high quality. The intent is to cover the theory, applications, and perspectives on the state of the art and future developments relevant to systems, decision making, control, complex processes and related areas, as embedded in the fields of engineering, computer science, physics, economics, social and life sciences, as well as the paradigms and methodologies behind them. The series contains monographs, textbooks, lecture notes and edited volumes in systems, decision making and control spanning the areas of Cyber-Physical Systems, Autonomous Systems, Sensor Networks, Control Systems, Energy Systems, Automotive Systems, Biological Systems, Vehicular Networking and Connected Vehicles, Aerospace Systems, Automation, Manufacturing, Smart Grids, Nonlinear Systems, Power Systems, Robotics, Social Systems, Economic Systems and other. Of particular value to both the contributors and the readership are the short publication timeframe and the world-wide distribution and exposure which enable both a wide and rapid dissemination of research output.

Indexed by SCOPUS, DBLP, WTI Frankfurt eG, zbMATH, SCImago.

All books published in the series are submitted for consideration in Web of Science.

Jaya Sil · Arup Kumar Bhaumik · Sonali Sen

Intelligent Real-Time Control Systems to Predict the Particle Size of Minerals Using Acoustics of Ball Mill

Springer

Jaya Sil
Department of Computer Science
and Technology
Indian Institute of Engineering Science
and Technology
Shibpur, Howrah, West Bengal, India

Arup Kumar Bhaumik
Camellia Institute of Engineering
and Technology
Budbud, By Pass(N), Burdwan
West Bengal, India

Sonali Sen
Department of Computer Science
St. Xavier's College (Autonomous)
Kolkata, West Bengal, India

ISSN 2198-4182 ISSN 2198-4190 (electronic)
Studies in Systems, Decision and Control
ISBN 978-3-031-91289-4 ISBN 978-3-031-91290-0 (eBook)
https://doi.org/10.1007/978-3-031-91290-0

This Springer imprint is published by the registered company Springer Nature Switzerland AG
The registered company address is: Gewerbestrasse 11, 6330 Cham, Switzerland

Foreword

It is a great pleasure for me to write about the work done by Arup, one of my classmates, along with his guide Dr. Jaya Sil and student Dr. Sonali Sen, on a unique research project aimed at incorporating intelligence into conventional mining machinery, which is currently operated manually. Research in this area has been ongoing for a long time, and I am very happy to note that the entire work has been included in a book. The brief of the book is as follows:

In today's fast-evolving industrial landscape, real-time process monitoring and control are of paramount importance. Intelligent systems are being rapidly integrated across various sectors, from plant operations to health care, revolutionizing traditional processes. This trend is particularly critical in mineral processing, where efficiency, precision, robustness, and sustainability drive operational success. Among the many challenges in the domain, accurately predicting the particle size distribution of minerals during crushing in a tumbling mill is a key area of focus, with the objective of minimizing mineral loss. This book, "*Intelligent Real-Time Control Systems to Predict the Particle Size of Minerals Using Acoustics of Ball Mill*", introduces an innovative approach that leverages the power of acoustics to redefine process control and optimization.

Acoustic analysis offers a non-invasive, highly responsive method for assessing mill performance. Unlike conventional techniques, which often involve delays, high costs, or invasive instrumentation, acoustic monitoring provides real-time insights into particle behaviour, mill load, and grinding efficiency. By integrating intelligent control systems with acoustic signals, this book presents a ground breaking framework that enhances predictive capabilities, minimizes operational variability, and ensures optimal milling conditions.

This work is not just about improving efficiency—it represents a shift toward smarter, more adaptive industrial processes. The fusion of acoustics, soft computing, and real-time control frameworks marks a new frontier in mineral processing, enabling industries to reduce energy consumption, improve throughput, and enhance overall productivity.

For researchers, engineers, and professionals in the field, this book serves as both a foundational resource and a guide. It offers theoretical insights, practical applications, and real-world case studies, equipping readers with the knowledge to implement acoustic-based control strategies in their own work. As industries move toward automation and intelligent decision-making, the principles outlined in this book will play a crucial role in shaping the future of mineral processing.

By bridging the gap between traditional methodologies and modern innovations, this book stands as a testament to the power of interdisciplinary research. The insights presented here will not only benefit academia and industry but also contribute to the broader goal of sustainable and efficient resource utilization. I wish all the best for the success of the book.

Dr. Ujjwal Maulik
Fellow IEEE, INAE, IAPR, AAIA,
NASI, Humboldt Fellow, Germany,
Fulbright Fellow, USA, Senior
Associate, ICTP, Italy
Professor and Former Chair,
Department of Computer Science
and Engineering
Jadavpur University
Kolkata, India
https://sites.google.com/site/
drujjwalmaulik/

Preface

In the mining industry, valuable minerals are extracted from the earth. It provides essential raw materials like coal, metals, and gemstones, which are used in construction, manufacturing, energy production, chemical and various other industries. To extract minerals from earth, there are several steps to perform. Ultimately when the minerals are available, the ore moves into the crushing process, where it is reduced in size for further processing, grinding, and mineral extraction. In the mining industry, the tumbling mills play a crucial role in the grinding and processing of ores. Ball mill is a kind of tumbling mill used to crush and grind the extracted minerals into finer particles such that valuable metals are easily separated from waste material during mineral processing. Ball Mill is a type of tumbling mill which is very commonly used for crushing and grinding. Ball mills operate by rotating a cylindrical shell filled with grinding media made of steel, iron or ceramic balls, which impact and grind the ore as the ball mill rotates. This process is essential for refining raw materials like gold, copper, and iron ore before further extraction methods such as flotation or leaching are applied. Efficiency, durability, and the ability are the essential characteristics of the ball mill to handle large volumes of particles, in mineral processing operations.

The working principle of the ball mill is actually based on impact and attrition. The raw ores are grinded into particles of different sizes depending on the requirements of the users with an objective to minimize the wastage of valuable minerals. The grinding procedure refers to the process of breaking down, sizing and separating the aggregate material. The ores should be grinded sufficiently fine so that the valuable minerals can be separated from the mixture of raw materials for further use. Experimental procedure to obtain a stipulated optimum particle size is critically lengthy and costly manual procedure. Generation of desired particle size depends on different parameters like initial size of raw ores, size distribution of the input mix, type and size of grinding media, type of grinding machines, speed of the machine, which are difficult to presume accurately without conducting experiments. Primary goal of a grinding machine is to extract maximum essential product from the given mixture of raw materials depending on the above-mentioned parameters. The major problem with this type of ball mill is that due to closed-loop operation there is a possibility of wasting a huge amount of raw materials. The motor of the ball mill is allowed to rotate, and the crushing process continues for a certain residence time. After that the motor is stopped and opening the lid, removing the media, the crushed particles are fed into a stack of screens to classify different size ranges. The desired particle size ranges are collected and rest of the materials, whose sizes are larger, are fed again for further dozing and lower sizes compared to the desired particle size are stacked in the waste stack. This process continues till to find two stacks of material; one is for desired output and another for wastage. This is a manual procedure where a huge amount of undersized raw material is produced, which is nothing but wastage. Also the efficiency of the procedure mainly depends on the efficiency of the human being who controls the system. The user cannot get any idea about the crushed product until or unless the mill becomes stopped and bring out the material for screening. However, it is difficult to accurately measure the time required to grind an input mix to its desired size. Raw materials are very much expensive, and this type of wastage is not a desirable proposition for a big industry. In spite of different past and recent studies there is no formula or model present, which can define the relationship between the given input and the expected output and can operate the mill from outside. Moreover, grinding parameters and residence time vary from minerals to minerals, and due to the absence of any such formula, one set of parameters, which are experimentally obtained, is not useful for another material, therefore not obtained without experimentation.

In the mining industry due to the large usage of ball mills, several researchers have studied this domain over the last few decades to control these types of mills automatically. Researchers tried to suggest several mathematical relationships, which to some extent can forecast how the broken particle varies with different ranges of particle size. However, none of them was good enough to interpret the results at coarse or fine particle size range. Moreover, the existing equations are not suitable enough to fit in an autonomous system where control can be imparted using information transaction only. During crushing the raw materials generate huge dust due to collision in between the grinding media and the materials, which prevent installation of sensors inside the vessel to capture information. Even though sensors are installed, they don't provide accurate results due to huge noise.

At the time of working with a ball mill in its load varying conditions in a noise free environment, it can be easily noticed that the acoustics of the ball mill varied gradually with crushing operation. The change is significant and one can easily understand that the particles are grinded, producing larger to smaller particles. So, acoustic is chosen as the main control parameter for designing the autonomous system in order to predict the different stages of grinding and correlate that with particle size distribution of the input mix. In reference to the working ball mill's acoustics, it is decided to install microphones outside of the ball mill for capturing signals during grinding operations. The acquired signal is preprocessed and analysed using computational methods to learn the acoustic signatures or patterns at different stages of grinding for various raw materials. In the loaded condition, the mill was allowed to run continuously to crush the material and sound was captured using the arrangement. The sound of the empty mill with only the grinding media is also captured to distinguish the empty ball mill from the loaded one. The acoustic signatures are stored along with corresponding particle size distribution and used to predict the desired particle size distribution for a new feed size and input mix at different stages of grinding. The recorded acoustic is stored in "WAV" file format. The size of the data stored for a loaded mill is approximately 10,000 K, with a varied range for a specific input mix to crush into its fine within 10 min. For an empty mill, the size of the signal is around 80–90 K for a few seconds. To make analysis simple, the continuous captured acoustics are segmented based on full mill cycle and transition cycle that fits within a specific scale maintaining continuity. To analyse the sound signal, different soft computing methods with mathematical modeling are proposed and finally validate the simulated results with the experimental outcomes. This book covers all those details including problems of the methods present at all. This book is divided into eight chapters.

Chapter 1 presents the overview of the crushing and grinding operation in the mining industry. The discussion starts with the grinding procedure and working principle of the ball mill. Next, the role of Acoustic Signatures in predicting the state of grinding using ball mill has been discussed. Then the problem statement is framed including the present research status of the grinding mechanism and the tools and techniques employed to develop the system. The book highlighted the new approach that has been presented towards solving the problems along with the implementation issues and challenges. Finally, the background studies involve different methods for Data analysis such as Data preprocessing, Data Organization, Partitioning of Data, Neural Networks for function approximation and Genetic Algorithm for predicting optimum particle size distribution. Petri Net representations are included in this chapter to represent the knowledge base of the operational functioning of the ball mill using If-Then rules, to design the intelligent control system.

Chapter 2 is dedicated to understanding the acoustics of the running ball mill which has been analysed in order to predict and model the empty and loaded condition of the running ball mill. The acoustics of the running ball mill in its different load varying conditions are captured using sensors, explained in this chapter. Different procedures related to handling the voluminous acoustic data are also discussed. The procedure to differentiate the acoustics of the loaded ball mill from the empty ball mill using Mel Frequency Cepstrum Coefficients (MFCC) has been presented here.

Chapter 3 has mainly focused on particle breakage modelling. This chapter aims to describe the development stages of an appropriate control schematic for dynamic breakage concerning particle size distribution. Experimental analysis of the data reveals some interesting results that help to predict the particle size distribution of materials obtained from grinding. A '*K*' factor is derived to establish a relationship of grinding parameters considering different ores. Artificial neural network is used in the chapter to obtain a desired output product distribution by learning the patterns of input mix and its corresponding particle size distribution, which are obtained experimentally. Experimental data are verified with simulation results for test input patterns to establish the predictive grinding method, using the proposed neural network-based approach.

Chapter 4 proposes two efficient strategies to predict the states of running ball mills. A Mathematical model-based approach is discussed to model the acoustic patterns generated by using the kernel density estimation (KDE) function applied on raw acoustic signals. The unique patterns are simulated using Gaussian curve fitting algorithm to smoothen the patterns, and the corresponding parameters are extracted to develop the model. In another approach, first statistical features are extracted, and then using principal component analysis (PCA)-based feature selection method, important features are selected. The data points with the selected features are clustered using K-means clustering algorithm, and Naive Bayes classifier is applied to predict the current running state of the ball mill for test patterns, nearest to that of a particular cluster, measured using Euclidean distance.

Chapter 5 proposes a novel technique to predict the stopping criteria of the mill with reference to the residence time of breakage of the materials in order to achieve the desired particle size distribution (*psd*). The proposed approach is performed in two different steps based on the information of ball mill acoustics. Using Genetic Algorithm, the desired state of grinding of the ball mill is predicted based on its two previous consecutive states, and after running the ball mill up to that predicted state, it is stopped automatically. In the second part, Gaussian Mixture Model and Fuzzy C-Means (GMM-FCM) clustering algorithm is applied to estimate the *psd* of the state at which the ball mill is stopped. When the ball mill stops, the *psd* of the breakage material corresponding to a particular state must belong to the respective cluster, verifying the experimental result.

Chapter 6 proposes a method to predict the particle size distribution in real time of a ball mill with specific input mix and fixed number of grinding media. This approach is based on the analysis of the time series data and validated using machine learning techniques. The proposed method is very easy to implement, and since no such complex transformations are used on the raw data, it is less time consuming. By this proposed method, the machine is stopped automatically after running the predefined residence time and corresponding desired particle size distribution is achieved.

Chapter7 demonstrates how Petri Net has been used successfully for implementing, monitoring, and controlling the activity of a tumbling mill to achieve optimum grinding within residence time. To simulate the entire scheme of a ball mill, the rule base has been designed and realized by an extended-Petri-net (EPN) structure. The operational status of the ball mill is explained using the rule-based reasoning process. The inferences generated from the reasoning process are converted into control actions to achieve the desired product in a real-time scenario. The expert control system is not only valid for ball mills but also equally working efficiently with other different kinds of reactors/crushers/compound machines, by modifying the rule base.

Chapter 8 concludes the book by summarizing the research objectives and indicating the directions for the future work.

Kolkata, West Bengal, India
March 2025

Jaya Sil
Sonali Sen
Arup Kumar Bhaumik

Competing Interests The authors have no competing interests to declare that are relevant to the content of this manuscript.

Contents

Abbreviations

ADC	Analogue-to-Digital Converter
AG	Autogenous
ANN	Artificial Neural Network
AR	Auto Regressive
BM	Ball Mill
BP	Back-Propagation
CNN	Convolution Neural Networks
DAC	Digital-to-Analogue Converter
dB	Decibels
DCT	Discrete Cosine Transform
DFT	Discrete Fourier Transform
DSP	Digital Signal Processing
EM	Expectation Maximization
EPN	Extended-Petri-Net
FCC	Forward Computation Cycle
FCM	Fuzzy C-Means
FFT	Fast Fourier Transform
FN	False Negative
FP	False Positive
FR	Frequency Range
GA	Genetic Algorithms
GMM	Gaussian Mixture Model
Hz	Hertz
IFT	Inverse Fourier Transform
KDE	Kernel Density Estimator
MFCC	Mel Frequency Cepstrum Coefficients
MLE	Maximum Likelihood Estimation
MLP	Multilayer Perceptron
NB	Naive Bayes
NN	Neural Network
P.E	Processing Element

PCA	Principal Component analysis
PDF	Probability Density Function
PSD	Particle Size Distribution
ROC	Receiver Operating Characteristics
SAG	Semi-Autogenous
SSE	Sum Squared Error
TN	True Negative
TP	True Positive
TSP	Traveling Salesman Problem

List of Figures

List of Tables

Chapter 1
Introduction and Preliminaries

Abstract This Chapter introduces the domain of the work where the grinding procedure of the ball mill and the role of acoustic signatures in predicting the state of grinding are discussed. After that, discussions are continued towards the problem statement, the present status of the grinding mechanism, and the tools and techniques employed to develop the system. Next the new approach has been presented towards solving the problem along with the implementation issues. The background studies involving methods such as Data preprocessing, Data Organization, Partitioning of Data, Neural Networks for function approximation and Genetic Algorithm for predicting optimum particle size distribution are included in this Chapter for better understanding of the materials and methods, applied for finding solution of the problem.

1.1 Introduction

Almost all the inorganic materials used to support our civilization are in some way derived from the earth's crust, available at the thin upper portion of the planet to the maximum depth of 20 K.M. The various elements that contribute to the development of the crust are not evenly distributed, rather they exist as a mixture of minerals. Minerals, by definition, have a definite formula in contrast to rock. Today, and more importantly in the future, the mined-out products are not suitable to be reckoned as final products in any industry. Therefore, the need for mineral processing [1, 2] has been felt and the researchers working with an objective to minimize the wastage of minerals. One of the important areas of mineral processing is size reduction. In terms of size and sometimes shape, commercial products need to meet some specific requirements in order to reduce the bulk of fibrous materials for easier handling.

In the mining industry several researchers have studied this domain over the last few decades. Due to the static control of the ball mill, which is manual, material wastage becomes a common phenomena. Researchers Rosin-Rammler, Gaudin-Schuman, Benette & Colleman [3–5] worked in this area and suggested several mathematical relationships, which to some extent can forecast how the broken particle

J. Sil et al., *Intelligent Real-Time Control Systems to Predict the Particle Size of Minerals Using Acoustics of Ball Mill*, Studies in Systems, Decision and Control 600,
https://doi.org/10.1007/978-3-031-91290-0_1

varies with different ranges of sizes. However, none of them were good enough to interpret the results at a coarse or fine particle size range while grinding. Moreover, the existing equations are not suitable enough to develop an autonomous system where control can be imparted using information transactions only.

During crushing the raw materials, huge dust is generated due to collisions in between grinding media and materials. Though for automation, sensors are very much helpful to acquire information, however in this case, none of the sensors like optical, magnetic or electric can be implanted within the vessel for dust, collision etc. Moreover, due to the dust generated by the grinding media, the prediction of particle size distribution using the sensor data is highly inaccurate.

The goal is to design an automatic control system that controls the operations of the ball mill in a closed loop environment resulting in minimum wastage of the valuable minerals. The approach is data driven therefore, through experimentation using the ball mill, data is captured in a noise-free environment at the University laboratory. It has been noticed that the acoustics of the ball mill gradually vary with particle crushing. The change is so significant that one can easily understand that the crushing is going on and particle size distribution changes with grinding depending on the variation of different parameters of the ball mill and grinding media. Moreover, based on the feed size or feed amount of the raw material, there is a noticeable difference in crushing, which the human ear can easily identify. In reference to the working ball mill's acoustics, it has been understood that the only sensor that can be used to capture the ball mill sound is the microphones or sound sensors that could be installed outside of the vessel. To predict the crushing stages of a ball mill operation, the acoustics of a running ball mill in different phases of grinding is analyzed to derive the signatures of the signal with different raw materials. After analyzing the acoustics using different methods, a model is proposed using which the state of operation of the ball mill in different load varying conditions with different input raw materials are predicted automatically. In a particular state of operation, the desired particle size is obtained after crushing with the suggested conditions.

1.1.1 Background

Before any research truly takes shape, there is an essential phase of preparation known as the 'design preliminaries,' where groundwork is laid for a successful outcome. Design preliminaries are the foundational steps in the design process, where designers clarify goals, constraints, and user needs, ensuring that the end product is both functional and user-centred. This phase is crucial because it helps to identify potential challenges early on, aligns all stakeholders on the vision, and provides a roadmap that guides the entire process. By investing time in the preliminary steps, designers can create solutions that are well-suited to user needs and adaptable to the different constraints specific to the applications.

Two main operations need to be understood are crushing and grinding, along with involvement of the corresponding devices. Crushers and grinders are used for size reduction (also known as comminution) in different industries. An ideal crusher or a grinder must have a large capacity to hold the raw material and consume as little energy as possible. At the same time the product particles must satisfy the requirements of different industries in terms of size, separation of unwanted large impurities from the input feed and others. In practice, all these criteria are hardly met by a single grinder or a crusher. The actual capacity of any crusher or grinder is always several times less than its ideal capacity and it is common to have products with a very large particle size distribution. Also, the power required for crushing or grinding is much higher than the calculated theoretical power. Sometimes, a screen can be put at the outlet of the crusher or grinder to remove the coarsest particles in the product, but this provision cannot eliminate the chance of filtering too-fine particles produced during crushing. Size reduction is the process of reducing large solid raw units or substances into smaller unit mass, coarse particles, or fine particles. The process of size reduction is also termed comminution or diminution or pulverization. This can be achieved by two methods:

(1) Precipitation method—the substance is dissolved in an appropriate solvent.
(2) Mechanical Method—the mechanical force is introduced by using different equipment.

1.1.1.1 Crushing and Grinding

Generally, size reduction is the combination of two different operations—Crushing and Grinding [6]. Industrial Size Reduction follows the mechanical method. Equipment used for grinding or crushing is divided into four categories—Crushers, Grinders, Ultrafine grinders and Cutting machines. Crushers do the heavy work of breaking large pieces of solids into smaller lumps. Primary crushers are run on materials out of mine, where the size may reach more than one meter and break it into 150–300 mm lumps. A secondary crusher reduces those lumps to particles in the size range of 5–10 mm. Grinders reduce the crushed feed to powder. The product from an intermediate grinder usually passes a 40-mesh screen (about 0.35 mm), while most of the product of a fine grinder passes a 200-mesh screen (0.074 mm). Screening is required to separate different-sized products produced at the end of the grinding process. Ultrafine grinders accept a feed of a few mm in size and reduce it to a product of sizes in the range of 1–50 μm. Cutters will deliver particles of definite shape and size, typically a few mm in length. In the working domain, several different types of size-reduction machines are available. Among them, the commonly used types of size reduction machines are:

- Crushers:
  - Jaw crusher
  - Impact crushers
  - Gyratory crushers
  - Crushing rolls

- Grinders:
  - Hammer mills and other impact types
  - Ball mills, Tube mills and Rod mills.
  - Roller mills and bowl mills
  - Attrition mills
- Ultrafine grinders:
  - Hammer mills with internal classification
  - Fluid energy mills
  - Agitated mills
- Cutting machines:
  - Knife cutters.

Grinding

The primary goal of grinding is to reduce the particle size of ore materials to facilitate the liberation of valuable minerals from the surrounding waste (gangue) material. This size reduction process is essential for achieving the required particle size for downstream processes like flotation, gravity separation, or leaching. During grinding, the particles can produce fine particles and the finer particles have a greater surface area, which improves the efficiency of subsequent recovery processes.

For grinding, various types of raw materials use different types of grinding mills and the choice of equipment depends on the factors like ore type, desired product size, and throughput. The following are the types of mills:

- Ball Mills: One of the most commonly used grinding machines for mineral processing. It consists of a rotating vessel filled with steel balls or other grinding media that reduce particle size through impact and abrasion. This vessel rotates on its horizontal axis.
- Rod Mills: Similar to ball mills but with steel rods instead of balls as the grinding media. Rod mills are used in coarser grinding and can produce slightly finer output than crushers.
- Autogenous (AG) Mills: In an AG mill, the ore itself is used as the grinding media. These mills can be used for coarse or intermediate grinding.
- Semi-Autogenous (SAG) Mills: A combination of autogenous and ball milling, where a portion of the grinding media is made up of steel balls and the rest is the ore itself.
- Vertical Mills: Used for fine grinding, especially when high surface area is required. They use vertical pressure and gravity to cause grinding.
- Stirred Mills: High-speed rotating stirrers are used to increase the grinding action, offering more efficient grinding for fine or ultra-fine material.

1.1.1.2 Grinding Media

The grinding media [7] plays a significant role in grinding the raw material. The choice of material, size, and shape of the grinding media affects the grinding efficiency and energy consumption. Grinding media can be chosen based on the types of raw materials to be grind. Steel Balls, one of the types of grinding media used generally in ball mills and SAG mills for their high density and hardness, providing a strong impact. For rod mills rods are used for producing a more uniform particle size distribution. Ceramic or Glass Balls are also used for finer grinding or for grinding sensitive materials that shouldn't be contaminated by metal. So based on different raw materials to crush and different types of mill, the grinding media should be decided.

1.1.1.3 Factors of the Grinding Process

Several factors influence the efficiency of the grinding process, like Ore Hardness, Feed Size, Media Size and type, Mill Speed, Grinding time and Water or chemical additives, are a few of them. The final product size and mineral liberation depend on the amount of time the mineral spends in the mill. The grinding process is an energy-intensive process. The energy efficiency can be affected by different factors. Feed size distribution is one of the key factors for energy efficiency. A more uniform particle size distribution reduces the grinding effort. Efficient media can reduce the amount of energy needed for grinding. So, the selection of grinding media is also an important feature for the energy efficiency of the grinding operation. Optimising mill speed, grinding time, and other variables also can reduce energy consumption.

1.1.2 Ball Mill

Ball mill is a type of grinder commonly employed in the mining and mineral processing industries. It is used to grind materials such as ores, chemicals, and ceramic raw materials to fine powders. The ball mill operates on the principle of impact and abrasion, where the grinding media (typically steel balls) inside the rotating mill shell collide with the ore, breaking it into finer particles. This type of grinding mill is versatile, operating in both wet and dry conditions, and can handle a wide range of feed materials, from coarse to fine. Throughout this book, "ball mill" may also be referred to simply as "mill".

Ball mill is often employed in the primary or secondary stages of ore processing to achieve the desired particle size for optimal mineral liberation. The size and composition of the grinding media, the mill's rotational speed, and the operational parameters (such as feed size, slurry density, and grinding time), which play important roles in determining the efficiency and effectiveness of the milling process. The working of a ball mill is based on impact and attrition: the impact, caused by the

Fig. 1.1 External view of a laboratory ball mill

balls dropping from the top of the shell, breaks down the particles, thereby resulting in a size reduction. Ball mill includes a hollow cylindrical shell which is rotated about its axis. The axis of the shell is either horizontal or slightly inclined, thereby making a small angle with the horizontal axis. The shell is partially filled with balls, which form the grinding medium of the mill. These balls are usually made up of steel, ceramic, flint pebbles, or hard rubber. The inner wall of the shell generally contains a coating of abrasion-resistant material, e.g., manganese steel or rubber. Rubber-lined (or coated) ball mills cause less wear to the products. The ball mill is nearly equal in length and diameter.

For grinding, it is filled with grinding media of different sizes and types depending on the impact required to grind the input ore mix loaded for crushing. The diameter of the grinding balls is 12 to 125 mm, and generally, balls occupy 30%-50% of the total mill volume. The rotating speed of the ball mill is 60 to 100 rpm. The external view of the ball mill is shown in Fig. 1.1 and the sectional view of the ball mill used in the laboratory is shown in Fig. 1.2.

During operation maximum 60% of the vessel volume can be filled with the input mix, otherwise, the material can leak from the vessel. The inside view of the working ball mill is given in Fig. 1.3.

1.1.2.1 Operation of Ball Mill

Different particle size ranges (larger than the desired output) can be mixed for the preparation of input by varying the speed of the dozer motor and also limiting the amount of dozing. A few balls are inserted inside the vessel with input materials and after tightening the lid of the vessel, the mill is allowed to start. The ball mill crushes the materials for a specified amount of time.

The crushing of materials in a ball mill is based on three important conditions:

- *Rotation speed of the mill*—Deciding the speed of the mill for crushing [8, 9] is very important. If the speed is too low, then the grinding media roll over on each

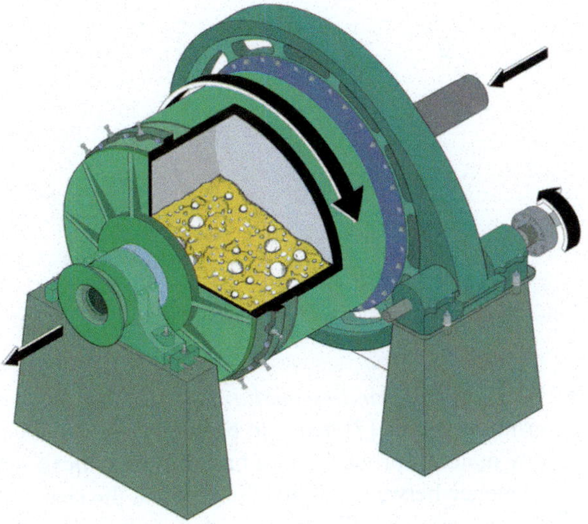

Fig. 1.2 A section of a ball mill

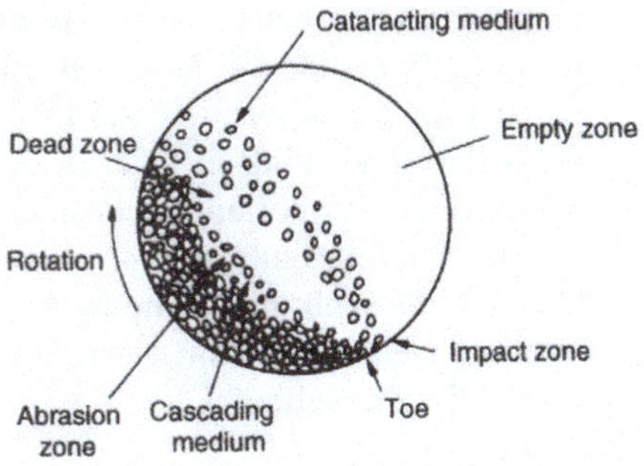

Fig. 1.3 Inside view of a working ball mill

other, which increases the attrition only. If the speed is too high, then both media and material fall on the wall of the cylinder. In both cases, it is not effective to reduce the size of the material. So, the decision to set up the correct speed of the ball mill for different input mixes and different size distributions of the raw material is very important for any crushing operation. The diagram of the internal working of the ball mill at different speeds is given in Fig. 1.4.

- Material residence time in the vessel: The ball mill is allowed to run for a specific period with a certain type of raw material. The material is then withdrawn from the mill and screened to check whether the desired particle size range is achieved or not. If not, continue the process by fixing again the specific time. However, due to arbitrarily fixing the time, the material may crush more than the desired size. Therefore, the crushed material cannot be used for further processing, which results in the wastage of the material. The decision of how much time the material will be in the ball mill for crushing is a very important parameter to decide.

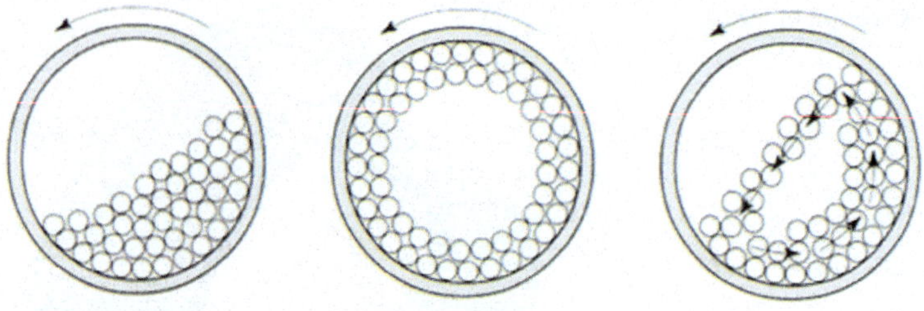

Fig. 1.4 Working of the ball mill at different speeds

- *Weight of the grinding media*: To crush the raw material, the correct weight of the grinding media [7] needs to be chosen so that it can crush the largest piece of the raw material properly. In a ball mill to achieve maximum operational efficiency, a balance between the grinding media, the size of the media and the feed material size is required. If the input mix contains many large lumps which cannot be crushed by the grinding media, then the situation leads to the gradual accumulation of materials between the balls. The situation may lead to operational suspension of the mill. To handle this situation either need to increase the media size or reduce the size of the input mix. An increase in the media size results in decreasing the ball mill-working surface, as a consequence the mill productivity is reduced. Moreover, it's important to consider the portion of the vessel, which will be filled by the grinding balls. Large filling leads to a situation where rising grinding balls collide with the falling balls. Based on the input particle mixture size, the grinding media composition, size, density, and hardness need to be chosen, which is discussed as follows
  - *Size*: For obtaining the smaller particle size [3] as the final product, smaller media particles are used. However, it must ensure that the grinding media should be large enough (i.e. more than 3–4 orders of magnitude) than the raw material size to be grinded.
  - *Density*: The density of the media should be high enough so that the material is grinded and does not float on top of the material.
  - *Hardness*: The media should be hard enough to grind the material, but where possible should not be so tough that it also wears down the vessel at a fast pace.
  - *Composition*: Different input mixes need different composition of the media. Steel, iron and ceramic balls depending on the type of input mix are used for this purpose. Grinding media, like iron, may react with corrosive materials. Despite iron-flint, stainless steel and ceramic grinding media can be used if corrosive substances are present for grinding.

1.1.2.2 Performance Criteria of Ball Mill

The performance of the ball mill is affected by several operating parameters of the mill. The mill speed is one of the primary criteria. The performance is also affected by the amount of grinding media inside the ball mill. The size of the incoming material also affects the grinding performance.

The speed at which the ball mill rotates is usually expressed as a percentage of the critical speed. It is the speed at which the centrifugal forces equal to the gravitational forces at the ball mill vessel's inside surface and no balls falling from its position onto the vessel. Therefore, at the critical speed, no further grinding is possible. Operating the ball mill too close to the critical speed can lead to inefficient grinding.

The performance of the ball mill depends on the grinding media. Too little media leads to ineffective grinding, while too much reduces the available space for material to be grind. The size and distribution of the grinding balls have a direct impact on the efficiency of the milling process. Larger balls are generally used for coarse grinding and smaller balls for finer grinding.

Input feed of material has a relationship with the grinding performance. Typically, the larger the feed, the coarser the product will produce. The concentration of water or other liquid in the ball mill (typically referred to as the pulp density) affects the grinding efficiency. If there's too much liquid, it can dilute the concentration of solids, reducing the effectiveness of the milling process. During certain grinding processes, heat can affect the performance of the ball mill, especially in mineral processing [1, 2, 6] where temperature-sensitive materials are used. In general, to obtain the desired particle size, the input mix is prepared with different particle size distributions using a dozer.

At runtime, the ball mill rotates around its horizontal axis. The principle of grinding is based on critical speed. In general, to obtain the desired particle size, the input mix is prepared with different particle size distributions using a dozer.

1.1.2.3 Dozer

A dozer is an essential piece of heavy equipment often used in industrial and manufacturing settings, including the preparation of input mixes for ball mills. The dozer plays a pivotal role in the handling, blending, and feeding of raw materials to ensure uniformity and efficiency in the milling process. The dozers can efficiently handle large volumes of material, which quickly helps to streamline the input mix preparation process. By carefully controlling the blending and movement of raw materials, the dozer ensures a consistent mix, improving the ball mill's performance. Dozers are highly reliable for continuous use in industrial settings since they are designed for operation in harsh conditions. Following are the functions of a Dozer for the Input Mix Preparation used in the ball mill.

- Material Handling—The dozer is responsible for managing raw materials such as limestone, clay, iron ore, and other additives. It moves these materials from storage areas, such as stockpiles or hoppers, to the mixing zones.
- Blending of Raw Materials—Proper mixing of raw materials is critical to ensure consistency in the ball mill's output. The dozer's blade is used to layer and mix various components, to achieve a homogenous blend. This step is vital to maintain the desired chemical composition and physical properties of the final product.
- Feed Regulation—The dozer controls the rate at which the prepared mix is fed into the ball mill. By adjusting the flow, it helps to prevent overloading or underfeeding, ensuring optimal grinding efficiency.
- Stockpile Management—Dozers organize and level stockpiles of raw materials, making it easier for loaders or conveyors to access and transport them to the ball mill. This activity helps minimize material segregation and ensures consistent input quality.
- Clean-Up Operations—Dozers are also used for cleaning spillage around conveyor belts, hoppers, or other equipment in the input mix preparation area. This keeps the production site safe and operational.

1.1.2.4 Usefulness of Ball Mill

The use of a Ball Mill is advantageous in different cases. Ball mills are easy to maintain and operate. Dry and wet, both types of grinding are possible using the ball mills, which proves its versatility in the real world. Ball mills can achieve fine particle sizes, which are necessary for mineral liberation in subsequent separation stages. The ability to operate in both continuous and batch modes makes them versatile for various processing plants. These types of mills can be used in small-scale operations or scaled up for large industrial operations. Using a ball mill may also be disadvantageous in certain cases. Ball mills are energy-hungry machines that require significant amounts of power to operate, especially when fine grinding is needed. The grinding media (balls) wear out over time, which can lead to contamination of the product and the need for regular maintenance. Because of the prolonged grinding time, ball mills can lead to over-grinding, where the desired particle size is exceeded, wasting energy and potentially reducing recovery rates in subsequent processes. Ball mills are more effective in reducing material size to a medium or fine particle size. For ultra-fine grinding, other mills (such as stirred mills or vertical mills) might be more efficient.

Due to cost-effectiveness, the ball mill is used for industrial purposes. In addition, it is very efficient to grind hard materials. After charging the ball mill with input mix and grinding media, the mill starts to run. The user cannot get any idea about the crushed product until or unless the ball mill stops and brings out the materials for screening. No one can accurately measure the time required to grind an input mix to its desired size. After crushing, when screening of the material takes place, then if it finds that the material size is larger than the required material size, the same method can also be applied once more and get the output. But if the material size becomes smaller than the required one, then the total material becomes waste. Raw materials are very expensive, and this type of waste is not desirable for a big industry. Despite different past and recent studies, no formula can define the relationship between the given input and the expected output, and the ball mill can be operated from outside.

1.1.3 Acoustics

The scientific study of sound is known as Acoustics. The study includes—how the sound is produced, transmitted, and received, as well as how it interacts with the environment. This part of our study encompasses a wide range of topics, including the physics of sound waves and the design of spaces optimised for sound. Different fields like engineering, music, architecture, medicine, and environmental science are different application fields of Acoustics.

1.1.3.1 Physics of Sound

Fundamentally, sound is a mechanical wave that travels through a medium, such as air, water, or solids. Sound is created if a vibrating object disturbs the surrounding medium and sets the particles of that medium into motion. A sound wave has several key physical characteristics. Sound travels in the form of longitudinal waves, where particles in the medium oscillate parallel to the direction of wave propagation. These waves consist of compressions (regions of higher pressure) and rarefactions (regions of lower pressure). The sound wave must have a frequency which determines its pitch and represents the number of vibrations (or cycles) per second. Frequency is measured in Hertz (Hz). The human hearing range is typically between 20 to 20,000 Hz. A sound must have an amplitude that determines its loudness. Amplitude is measured in decibels (dB). The human ear can sustain a sound with amplitude typically ranging around 60 to 70 dB. The wavelength of a sound wave can be extracted using the frequency of a sound wave. Distance between two consecutive compressions or rarefactions in a sound wave is named as wavelength that is inversely related to frequency; higher frequencies have shorter wavelengths and lower frequencies have longer wavelengths. A sound wave has a speed which depends on the medium through which the sound travels. At room temperature (20 °C), sound can travel through the air at approximately 343 m per second. Typically, sound can travel faster in water and even faster in solids. Acoustics is a fascinating and diverse field that touches almost every aspect of real life and plays a central role in shaping the world.

1.1.3.2 Acoustic Signature

An acoustic signature is the unique pattern of sound, vibration, or noise emitted by an object, system, or environment. The acoustic signature can be defined by various factors, such as frequency, amplitude, duration, and spatial characteristics of the sound waves. It can include both the frequencies and intensity of sounds, as well as their spatial characteristics (where and how the sound travels). In essence, an acoustic signature acts like an audible fingerprint—distinct and unique to each sound-producing entity. Much like how each human voice has a particular tone or timbre, objects and environments—ranging from mechanical systems to natural phenomena that generate a set of identifiable acoustic characteristics. These signatures can be used to distinguish one object or system from another, often providing critical insights into their identity, behaviour, or operational state.

1.1.3.3 Microphones

Analysis of acoustic signature needs different steps to follow, starting with capturing the sound using microphones or other sound sensors, preprocessing, converting the sound into electrical signals and finally taking decisions by analysing the signal. Microphones are the primary devices for capturing sound; other sensors also measure sound or acoustic signals. These sensors often operate in specific contexts or environments, such as industrial settings, or for scientific research.

A microphone is a type of sensor that converts acoustic energy into electrical signals. It is one of the most common sensors for capturing sound, and it's used in a wide range of applications, from recording audio to communication systems and environmental sound detection. A microphone has two components—the Diaphragm and the Transducer Mechanism. The diaphragm is a thin, flexible material that vibrates in response to sound waves. The transducer converts the diaphragm's movement into an electrical signal. Figure 1.5 is the internal representation of a microphone.

1.1.4 Signal Processing

Signal Processing refers to manipulating and analysing signals—such as sound, images, or other data. Signals can be of two types—Digital and Analog. Unlike Analog signals, which are continuous in time and amplitude, digital signals are discrete and represented in binary form. Digital Signal Processing (DSP) involves the conversion of real-world Analog signals into digital form using an analogue-to-digital converter (ADC), followed by various mathematical operations to process,

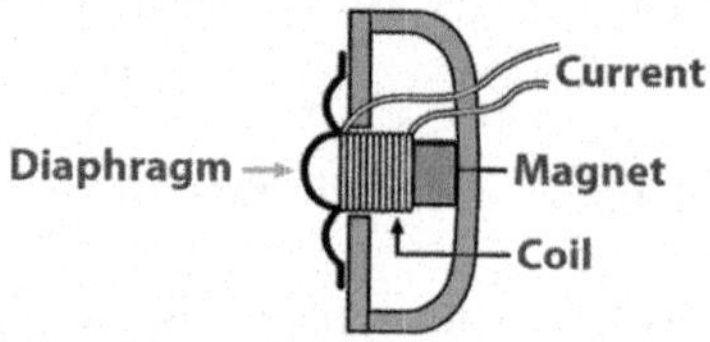

Fig. 1.5 Internal diagram of a microphone

enhance, or analyse the signal. The digital signal is converted back to analogue form using a digital-to-analogue converter (DAC).

Sound signal processing, referred to as audio signal processing, is a subset of Digital Signal Processing (DSP) that specifically focuses on analysing, manipulating, and transforming sound signals. These sound signals can represent voice, music, environmental sounds, or any other form of acoustic signal. Sound signal processing plays a crucial role in various applications, including audio recording, speech recognition, music production, communication systems, and hearing aids. Analog sound signals are continuous, representing fluctuations in air pressure (the sound waves, which can be hard). Digital sound signals, on the other hand, represent these fluctuations in a discrete format (samples of the continuous signal). Sampling and quantisation are used to convert an analogue sound signal to a digital form. The process results in a sequence of numbers (samples) that approximate the original signal. The sampling rate determines how often the analogue signal is sampled. Commonly used audio sampling rates include 44.1 kHz (used for CDs) and 48 kHz (used in professional audio and video systems). According to the Nyquist theorem, the sampling rate should be at least twice the highest frequency of the sound to avoid aliasing. The number of bits used to represent each sample in the digital domain affects the dynamic range and precision of the signal. More bits (e.g., 16-bit, 24-bit) allow more accurate representations of the amplitude of sound signals. The sound signal is represented as an amplitude versus time plot in Fig. 1.6 as the time domain representation. This is the raw representation when recording or analysing a sound wave in real-time. It provides information about the amplitude with respect to 0 and the timing of the sound, but it doesn't directly provide information about the frequency components. In some audio applications, the envelope of a sound signal (the smooth curve that outlines the peaks of the waveform), shown in Fig. 1.7 is more relevant than the waveform itself. Envelopes are used in tasks such as compression or noise gating, where the goal is to modify the loudness over time. When analysing or processing a segment of a sound signal, it is often helpful to break the signal into smaller pieces, or windows, before performing operations like Fourier analysis. This helps to minimize edge effects that can distort the analysis.

Audio signals are often analysed in the frequency domain by transforming the time domain signal into the frequency domain using Discrete Fourier Transform or DFT. This technique decomposes a sound signal into its constituent frequencies, revealing the spectral content of the sound. In the frequency domain, a signal is represented by its frequency components (e.g., tones, and harmonics), with each frequency component having a specific amplitude and phase. A signal may have a 3D representation of its spectrum over time known as a spectrogram. It shows how the spectral content of a sound changes with time and is commonly used in speech and music analysis. Sound in the frequency domain can be analysed for pitch (the perceived frequency) and harmonics (integer multiples of a fundamental frequency). For instance, a musical note may have a fundamental frequency and a series of harmonic frequencies that together define its timbre.

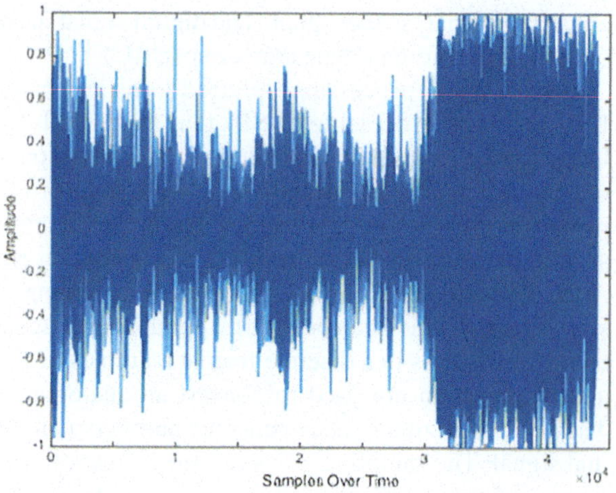

Fig. 1.6 Time domain representation

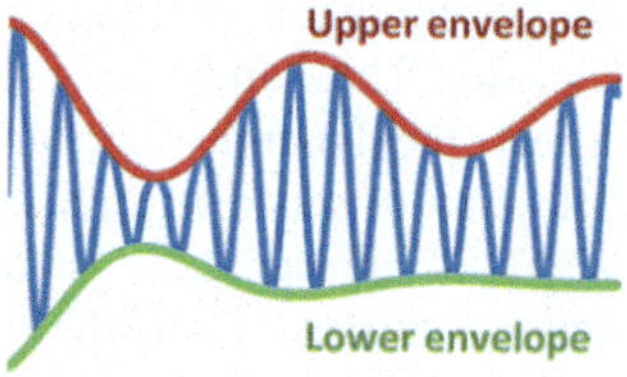

Fig. 1.7 Signal envelope

1.1.4.1 Signal Transformation

Transformation of the signal is just another form of representation of the original signal. It does not change the information content. An acoustic signal of the ball mill is converted into frequency domain data using Fast Fourier Transform (FFT) [10, 11] without any significant loss. The time domain acoustic signal of the ball mill is used to develop an autonomous control system. The data is represented in the form of a 1-D array in a monophonic recorder. In this case of a sound signal, for a five-second recording, the generated time domain continuous data is approximately fifty thousand, which is too large for manual analysis. The time domain data can be converted into frequency domain discrete data using transformations like the Discrete Fourier Transform (DFT) [12] and FFT. FFT is more effective than DFT since it decreases time complexity from N^2 to $N \log_2 N$ where N is the number of discrete signals and the implementation is much simpler. FFT follows the divide and conquer approach, where the original N point samples are divided into two ($N/2$) sequences. Figure 1.8 shows an example of the same.

The Fourier Transformation of a signal is represented as the sum of new transforms.

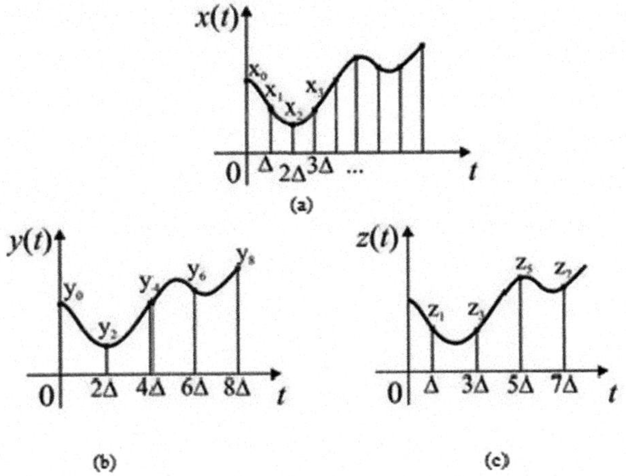

Fig. 1.8 Original signal (**a**) is divided into two new signals (**b**) and (**c**)

$$
\begin{aligned}
X_k &= \sum_{r=0}^{N-1} x_r e^{-i\frac{2\pi kr}{N}} = \sum_{r=0}^{N/2-1} x_{2r} e^{i\frac{2\pi k(2r)}{N}} + \sum_{r=0}^{N/2-1} x_{(2r+1)} e^{i\frac{2\pi k(2r+1)}{N}} \\
&= \sum_{r=0}^{N/2-1} x_{2r} e^{i\frac{2\pi k(2r)}{N/2}} + e^{-i\frac{2\pi k}{N}} \sum_{r=0}^{N/2-1} x_{(2r+1)} e^{i\frac{2\pi k(2r+1)}{N/2}}
\end{aligned} \tag{1.1}
$$

where r is the sample number, and integer k corresponds to the frequency over time of the wave.

There are two new Fourier Transform signals, given in Eq. (1.1). Two real and one complex variables are defined in Eqs. (1.2), (1.3) and (1.4), respectively.

$$Y_k = \sum_{r=0}^{N/2-1} x_{2r} e^{i\frac{2\pi kr}{N/2}} \tag{1.2}$$

$$Z_k = \sum_{r=0}^{N/2-1} x_{(2r+1)} e^{i\frac{2\pi kr}{N/2}} \tag{1.3}$$

$$W = e^{-i\frac{2\pi}{N}} \tag{1.4}$$

In reference to the Eqs. (1.2), (1.3) and (1.4), the Eq. (1.1) can be re-written as follows given in Eq. 1.5:

$$X_k = Y_k + W^k Z_k \tag{1.5}$$

FFT gives a clear view of the frequencies present in the signal and the proportion of the signal frequencies. Figure 1.9a depicts a random signal and the corresponding FFT representation is given in Fig. 1.9b.

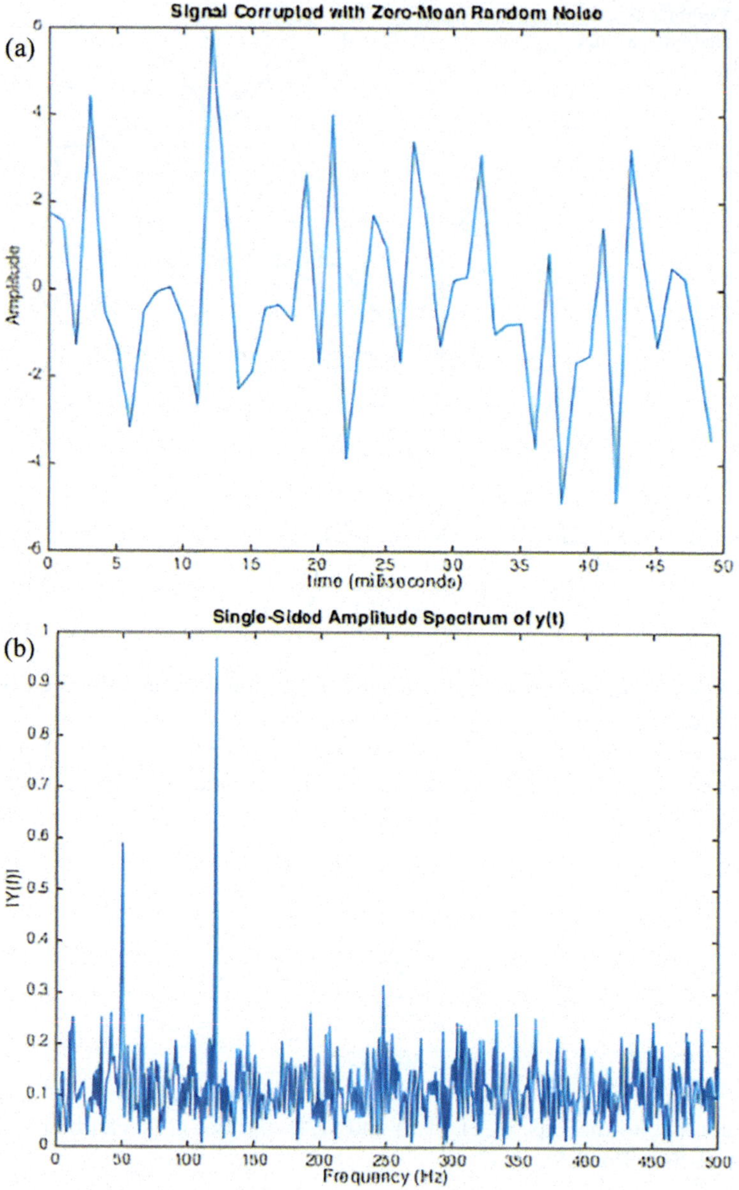

Fig. 1.9 **a** Time domain representation of a random signal. **b** Frequency domain representation of the random signal

1.1.4.2 Filtering

A sound may not always be captured in a noise-proof environment. Audio signals can be filtered to modify or enhance specific frequency components. Filtering is used to attenuate high-frequency noise or to isolate low-frequency components (e.g., bass sounds). Three very commonly used filtering methods—High-pass filter, Band-pass filter and Notch filter. The high-pass filter detects low-frequency rumbles or hums within the recordings. The Band-Pass filters are used to isolate a specific range of frequencies, often used in equalisers to enhance certain parts of the audio spectrum. Notch filters remove a narrow band of frequencies and are useful for eliminating specific noise sources, such as hum, from electrical interference. Filtering can also be used to shape the tone of an audio signal, for example, in audio mixing and sound design, by boosting or cutting frequencies within certain ranges. One of the most widely used techniques in sound signal processing is dynamic range compression. It reduces the difference between the loudest and quietest parts of an audio signal, making the sound more consistent. Sound signal processing is a diverse and rapidly advancing field that plays a central role in numerous applications ranging from music production and speech recognition to telecommunications and hearing aids. Sound engineers and researchers can enhance, modify, and understand audio signals in meaningful ways by employing a range of techniques such as filtering, time-domain manipulation, frequency-domain analysis, and dynamic range control. As technology continues to evolve, sound signal processing will remain a key development area, especially with advances in machine learning, real-time audio effects, and high-quality audio reproduction systems. Ultimately, by analyzing the sound signal, the acoustics signature can be generated. The generated pattern can then be analysed using different machine-learning techniques for modelling, verification and validation.

1.1.5 Feature Extraction Tool—MFCC

Feature extraction approaches usually yield a high dimensional feature vector after extracting information from the audio signal. Mel Frequency Cepstrum Coefficients (MFCC) [13, 14] is a very popular technique for feature extraction, used in the field of speech recognition, speaker recognition and audio classification. It is based on certain properties of the human auditory system, namely critical band frequency resolution and log power. Following are the steps to find the MFCC of a given acoustic signal:

(i) Pre-emphasis: The audio signal is passed through a high-pass filter and then performs the z transform on the filtered signal, as given in Eq. (1.6).

$$H(z) = 1 - az^{-1}, \text{ where } 0.95 < a < 0.99 \tag{1.6}$$

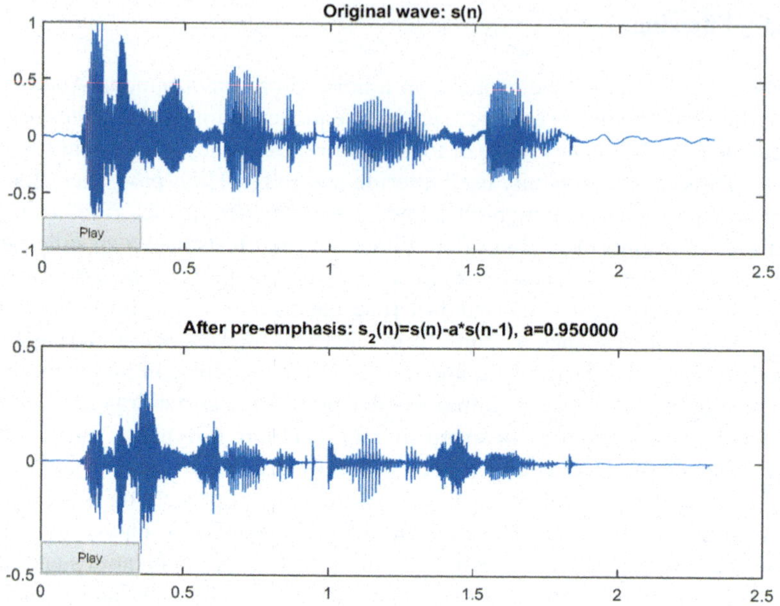

Fig. 1.10 Pre-emphasis signal of an original signal

The goal is to amplify the important high-frequency signal, which was suppressed during the sound production mechanism of humans. Figure 1.10 shows the pre-emphasised version of an original signal.

(ii) Frame Blocking: The output of the pre-emphasis step is segmented into frames of 20 to 30 ms with overlapping scenarios. If the frame size is not in the power of two, then zero padding is required to the nearest length of power two.
(iii) Hamming Window: This step maintains the continuity in between first and last points in the frame. The signal is denoted as $s(n), n = 0, 1, \ldots., N - 1$, and after applying Hamming window, the signal is represented as $s(n) * w(n)$, where $w(n)$ is the hamming window defined by Eq. (1.7).

$$w(n, \alpha) = (1 - \alpha) - \alpha \cos\cos\left(\frac{2\pi n}{N - 1}\right), 0 \leq n \leq N - 1. \quad (1.7)$$

Figure 1.11 shows the hamming window representation of the signal, given in Fig. 1.10.

(iv) FFT: The next step is to perform FFT to obtain the magnitude and frequency correlation of each frame.
(v) Mel-Scale Filter Conversion: Mel is a unit of pitch in which pairs of sounds are perceptually equidistant and are separated by an equal number of Mel. Mel-scale is approximately linear below 1kHz and logarithmic above 1kHz. The

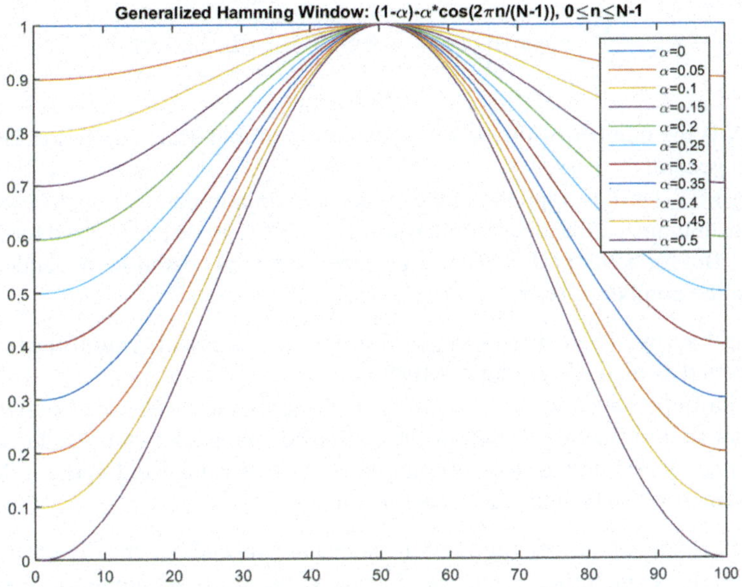

Fig. 1.11 Hamming window of the signal

FFT output is multiplied by 20 mel-scale frequency to get the log energy of each triangular band-pass filter. It is defined using Eq. (1.8), where f is the signal frequency.

$$Mel(f) = 2595 \log_{10}\left(1 + \frac{f}{700}\right) \tag{1.8}$$

Figure 1.12 shows the representation of the signal after mel-scale conversion.

(vi) Discrete Cosine Transform (DCT): DCT is applied on 20 log energy, obtained from the previous step, defined using Eq. (1.9).

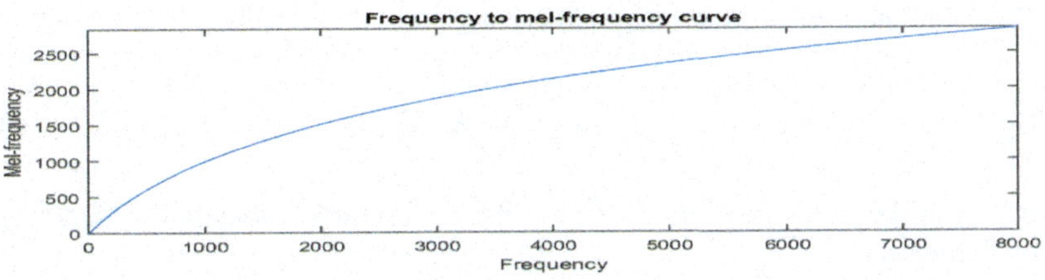

Fig. 1.12 Mel-frequency representation of the signal

$$C_m = \sum_{k=1}^{N} \cos\cos\left[m * (k - 0.5) * \frac{\pi}{N}\right] * E_k, m = 1, 2, \ldots, \tag{1.9}$$

where N is the number of filters, L is the number of mel-scale cepstral coefficients, while in general, $L = 12$.

After FFT, DCT is applied, so the frequency domain data is transformed into a time-like domain, called the quefrency domain. MFCC alone can be used for acoustic features. But for better performance log energy and delta operations are added which increase the number of features for analysis.

(a) Log Energy: The energy of each frame is an important feature that could be counted as the 13th feature to MFCC.
(b) Delta Cepstrum: Cepstrum is obtained by applying the Inverse Fourier Transform (IFT) of the logarithm of the estimated signal spectrum. Delta cepstrum can be represented as time derivatives of the cepstrum (i.e. Energy + MFCC), which also can be added as a new feature.

If the velocity and acceleration of the signal are also added, the dimension of the feature vector is increased up to 39. Generally, for acoustics, commonly available features are—amplitude, phase and frequency. In order to analyze the acoustics of the ball mill, enough features are required, which can be obtained using MFCC.

1.1.5.1 Feature Selection

Principal component analysis (PCA) [15] converts a set of correlated variables into a set of values of uncorrelated variables by using orthogonal transformation. It selects the first PC having the largest possible variance and each succeeding component turns under the constraint that it is orthogonal to the preceding components. The output vector is an uncorrelated orthogonal basis set. In PCA it transforms an $N \times d$ matrix X into an $N \times m$ matrix Y, where $m < d$.

The steps of PCA are described below:

Step 1: Centralize the data (i.e. Subtract the mean).

Step 2: Calculate $d \times d$ covariance matrix:

$$C = \frac{1}{N-1} X^T X \tag{1.10}$$

where $C_{i,i}$ is the variance of i and $C_{i,j}$ is the covariance between variables i, j and can be defined as

$$C_{i,j} = \frac{1}{N-1} \sum_{q=1}^{N} X_{q,i} X_{q,j} \tag{1.11}$$

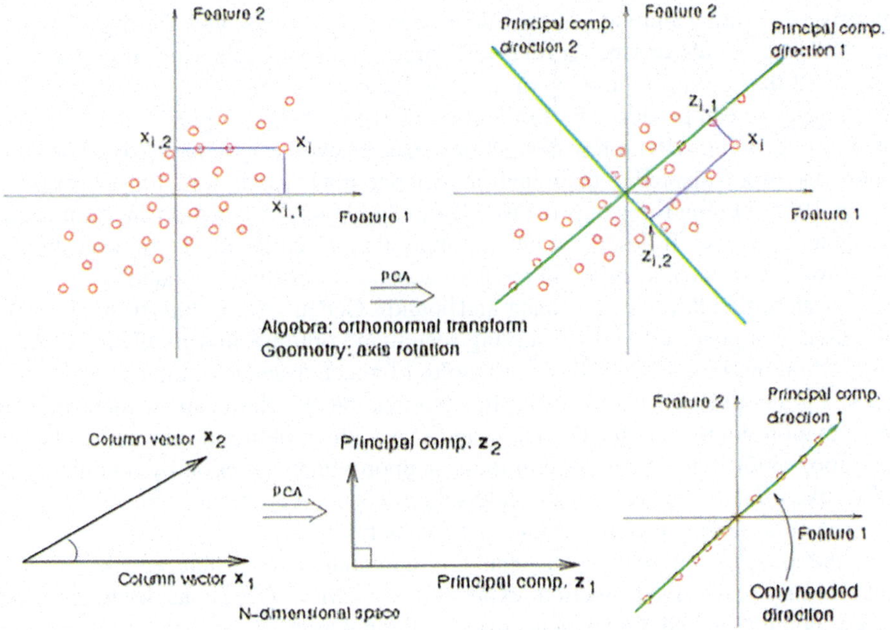

Fig. 1.13 1st PC and 2nd PC

Step 3: Compute eigenvectors $(e_1, e_2, \ldots\ldots .e_d)$ and the corresponding eigenvalues $(\lambda_1, \lambda_2, \ldots\ldots .\lambda_d)$.

Step 4: Sort the eigenvectors by decreasing eigenvalues and choose k eigenvectors with the largest eigenvalues to form a $d \times k$ dimensional matrix W.

Step 5: Use this $d \times k$ eigenvector matrix to transform the samples onto the new subspace, which can be summarized by the mathematical Equation $y = W^T \times x$, where x is a $d \times 1$ dimensional vector representing one sample and y is transformed $k \times 1$ dimensional sample in the new subspace.

To trace out the important features from a set of features, PCA is a good choice. However, it is a linear method used for feature selection. The first PC and second PC of an example are shown in Fig. 1.13.

1.1.6 Evolutionary Algorithm

Charles Darwin stated the theory of natural evolution as, "over several generations, biological organisms evolve based on the principle of natural selection 'survival of the fittest' to reach certain remarkable tasks". In nature, an individual in the population competes for basic resources such as food, shelter, etc. Also in the same species, individuals compete to attract mates for reproduction. In this selection procedure, poorly performing individuals have less chance to survive, and the most adapted or "fit" individuals produce a relatively large number of offspring. It can also be observed that during reproduction, a recombination of the good characteristics of each ancestor can produce "best fit" offspring whose fitness is greater than that of a parent in general. After a few generations, species evolve spontaneously to become more and more adapted to their environment and may sustain for a longer period.

In 1975, Holland developed this idea in his book "Adaptation in natural and artificial systems". It described how to apply the principles of natural evolution to optimize the problems and build the first Genetic Algorithms. Holland's theory has now been further developed by leaps and bounds. Genetic Algorithms (GAs) [16–18] are used as a powerful tool for solving search and optimization problems. Genetic algorithms are based on the basic principles of genetics and evolution.

GA is a population-based stochastic optimization algorithm, applicable when the information content is not too large and exact for applying a conventional optimization algorithm. The algorithm finds an approximate solution by optimizing an objective function because an absolute solution is not possible to achieve.

The optimization problem can be defined as follows:

Find $x_0 \in X$, in such a way that f is maximal in x_0, where $f : X \rightarrow R$ is an arbitrary real-valued function, expressed as $f(x_0) = f(x)$, considering x_0 is the initial population and X represents the set of populations.

1.1.6.1 Chromosomes

In GA the problem is formulated using Chromosome, which is a structure that contains genetic material and serves as a blueprint for the characteristics of an organism. Chromosome encoding is a key concept in genetic algorithms (GAs), where potential solutions to a problem are represented as chromosomes. These chromosomes are used to simulate the process of natural evolution, taking part in operations like selection, crossover, and mutation to find optimal or near-optimal solutions. The key elements of Chromosome Encoding are as follows:

(a) Representation: Each chromosome is a sequence of symbols (genes) representing a candidate solution. There are different techniques to encode a chromosome. Following are few of them

1. Binary Encoding: Uses binary strings (e.g., 101010).
2. Real-Number Encoding: Represents genes as floating-point numbers.
3. Permutation Encoding: Represents ordered lists, useful in problems like the Traveling Salesman Problem (TSP).
4. Tree Encoding: Uses tree structures, often applied in genetic programming.
5. Custom Encoding: Tailored to specific problem domains.

(b) Gene and Allele:

1. A gene is a single element of the chromosome (e.g., a bit in binary encoding).
2. An allele is a specific value a gene can take (e.g., 0 or 1 in binary encoding).

(c) Chromosome Length: It determines the complexity and resolution of the solution space. Longer chromosomes can represent more detailed solutions but increase computational complexity.

For example, say the objective is to select the minimum of the numbers from 39, 46, 89, 17 considered as the initial population and represented by the chromosomes of length seven. The binary encoding of chromosomes are given below.

p1: 0100111
p2: 0101110
p3: 1011001
p4: 0001001

1.1.6.2 Functions and Operations

The structure of evolutionary algorithm is based on the following components:

(a) Fitness function: In the biological evolution procedure, the fitness function is like the habitat to which the organisms adapt. It decides how the populations will change over generations. This function not only identifies the good or bad solution but also distinguishes a somewhat complete solution from a more complete solution.
(b) Population of chromosomes: Chromosomes refer to the numerical values of the parameters representing probable solutions to the problem. GA begins with randomly chosen chromosomes known as the initial population. At each generation, the chromosomes are evaluated by using the fitness function to test how far it is from the optimum solution.
(c) Selection Operator: The selection operator is used to choose the chromosomes for reproduction based on predefined probability, decided by the user. Selected chromosomes are used for further processing.
(d) Crossover Operator: The crossover operator is applied for information exchange between the pair of chromosomes and generates new off-spring, which are supposed to be better than the parent chromosomes. Crossover occurs during evolution according to a user-definable crossover probability. Crossover selects genes from the parent chromosomes and creates a new offspring, as shown in Fig. 1.14.

The Crossover operators are of many types and are selected based on the way chromosomes are encoded.

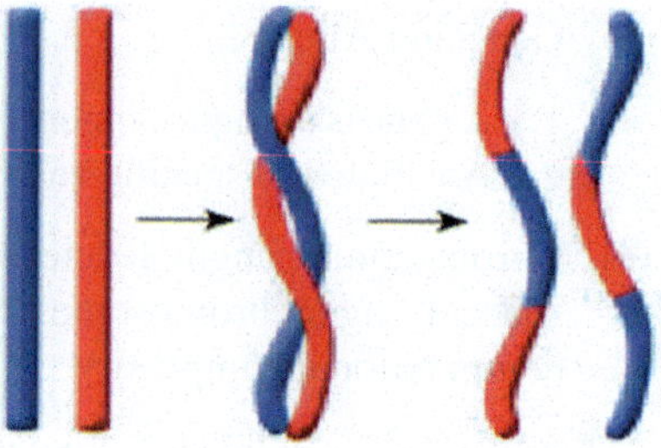

Fig. 1.14 Crossover operation

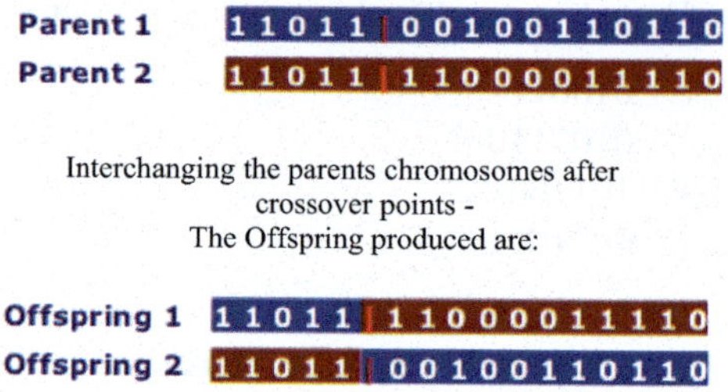

Fig. 1.15 One point crossover

- One-Point Crossover: This operator randomly selects one crossover point and then copy everything before this point from the first parent to the second parent and then everything after the crossover point from the second parent to the first parent. The Crossover would then look as Fig. 1.15.

  Consider the two parents selected for crossover.

- Two-Point Crossover: Two-Point crossover operator randomly selects two crossover points within a chromosome and then interchanges the two parent chromosomes between these points to produce two new offspring. Two-point crossover is shown in Fig. 1.16.
- Uniform Crossover: Uniform crossover operator decides (with a probability—known as the mixing ratio) which parent will contribute how much the gene values in the offspring chromosomes. The crossover operator allows the parent

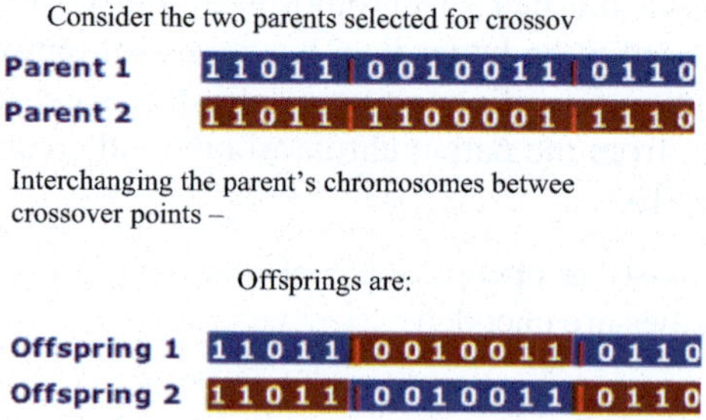

Fig. 1.16 Two-point crossover

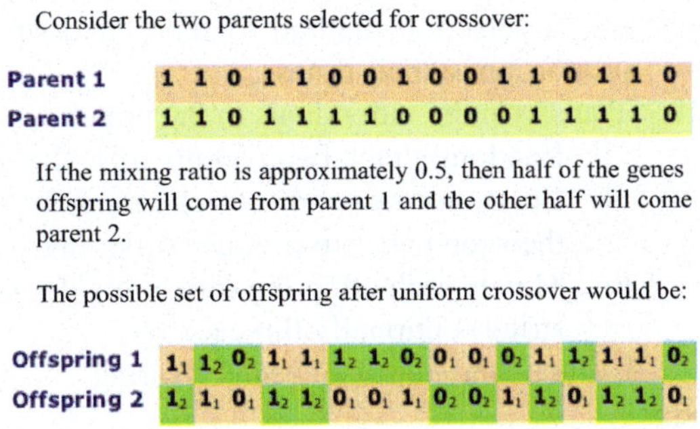

Fig. 1.17 Uniform crossover

chromosomes to be mixed at the gene level rather than the segment level (as with one and two-point crossover). An example is shown in Fig. 1.17.

(e) Mutation Operator:

Mutation Operator is applied to mutate the gene(s) through random deformation of the string of chromosomes with a certain probability, much less than crossover probability. Selection and crossover keep the information of the fitter chromosomes with respect to the current generation. So, this may cause the evolution to converge too quickly based on the local optimum based on the global optimum. Due to the use of a mutation operator, the algorithm diverges more slowly by maintaining the diversity in the population.

(f) Terminating condition:

GA is expected to produce good solutions, but might not find a perfect solution. GA terminates based on the following condition(s):

- After a pre-specified number of generations.
- When an individual solution reaches a pre-specified level of fitness.
- When the variation of individuals from one generation to the next reaches a pre-specified level of stability.

The algorithm is described below:

Input: Initial Population

Step 1: Initialize the population

//Initialize population sufficiently large, as representative of the whole
//search space.

Step 2: Evaluate the fitness of each population

Step 3: While the termination condition does not hold

begin

a. Replicate individuals based on their fitness

Use a selection method to reproduce strings in the next generation in proportion to their fitness.

b. Transform the individuals in the population
   i. Randomly pick two parents from the population
   ii. Crossover the parents (pick a random point in the strings and exchange their top parts) to produce offspring
   iii. Mutate each offspring (randomly decide whether to flip each bit in the string) optionally flip each bit)

c. Evaluate the fitness of each offspring

End

Output: An optimal solution

This evolution technique is a good choice for the prediction of an unknown from a limited input dataset and the optimized solution is achieved within finite time.

1.1.7 Data Distribution and Curve Fitting

(i) Gaussian Distribution [19–21]:

Random variables play important roles in every field of study. Data distribution functions are the techniques used for measuring and analyzing the statistical and probabilistic properties of the random variables. Different distribution functions are available in the literature. Gaussian distribution is known as Normal Distribution [22], a type of probability distribution technique used for real-valued random variables. The Probability Distribution function is defined using Eq. (1.12).

$$f\left(x;\mu,\sigma^2\right)=\frac{1}{\sigma\sqrt{2\pi}}e^{-\frac{1}{2}\left(\frac{x-\mu}{\sigma}\right)^2} \tag{1.12}$$

where μ is the mean and σ is the standard deviation of the distribution representing a symmetric bell-shaped curve. One example is given in Fig. 1.18.

(ii) Curve Fitting: The curve fitting algorithm [21, 23] captures the trend in the data by assigning a single function across the entire range of data. The Gaussian Function is a bell-shaped distribution around its center with a width that smoothly decreases as it moves away from its centers along the x-axis. Gaussian curve fitting can be mathematically defined in Eq. (1.13).

$$y(x)=Ae^{\frac{(x-\mu)^2}{2\sigma^2}} \tag{1.13}$$

where the controlling parameters are defined as follows:

A is the maximum amplitude,μ is the mean and σ is the standard deviation which controls the width of the curve along the x axis. Figure 1.19 shows a Gaussian curve fitting example.

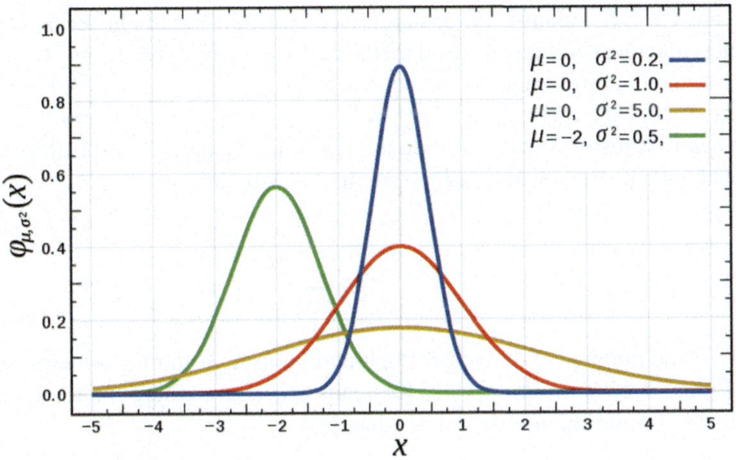

Fig. 1.18 Gaussian distribution

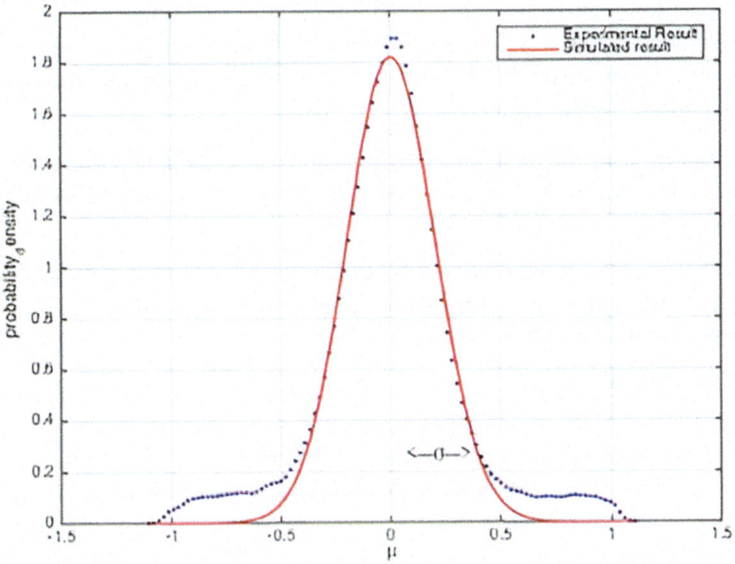

Fig. 1.19 Gaussian curve fitting

1.1.8 Clustering Algorithms [24]

The purpose of clustering is to divide the data into groups having similarities in a group and dissimilarities between different groups. It is an unsupervised learning technique. Following are some of the clustering methods that can be used during ball mill sound analysis are discussed below.

- K-means [25]: K-means is an unsupervised partition-based clustering algorithm, simplest and most commonly used method. The algorithm partitions the set of data into k clusters represented by the centers $C_1, C_2, \ldots\ldots\ldots, C_k$. An initial set of cluster centers are chosen randomly from the dataset. Each data point is assigned to its nearest cluster center based on the Euclidian distance and the cluster centers are evaluated as the mean of all the points in each cluster:

$$\mu_k = \frac{1}{N_k} \sum_{q=1}^{N_k} x_q \tag{1.14}$$

where N_k is the number of data points belong to cluster k and μ_k is the mean of that cluster. Figure 1.20 is an example of K-means procedure where the total number of data points is 50 and the number of clusters is 3.

- Fuzzy C-Means: Fuzzy C-Means [26] algorithm is a soft clustering technique applied to the cases where the clusters are not well separated as the objects belong to more than one cluster. It represents as collection of k clusters, $C_1, C_2, \ldots\ldots C_k$ and a partition matrix $W = w_{i,j} \in [0, 1]$ for $i = 1 \ldots .n$ and $j = 1 \ldots .k$, where each element $w_{i,j}$ is a weight that represents the degree of membership of belonging of an object i in cluster C_j. This method is iterative and used to approximate the minima of an objective function.

This algorithm allows data points to be assigned to more than one cluster and each data point has a degree of membership or probability of belonging to each cluster. The algorithm can be defined as follows:

Step 1: Randomly initialize the membership matrix so that $\sum_{j=1}^{C} \mu_j(x_i) = 1$; $i = 1, 2, \ldots ..k$ and μ_j is the degree of membership of x_i in the cluster j, c_j is the d-dimension center of each cluster.

Step 2: Calculate the centroid:

$$C_j = \frac{\sum_i [\mu_j(x_i)]^m x_i}{\sum_i [\mu_j(x_i)]^m}$$

Step 3: Calculate dissimilarity between the data points (x_i, y_i) and (x_j, y_j) centroid using the Euclidean distance $D_{ij} = \sqrt{(x_i - x_j)^2 + (y_i - y_j)^2}$.

Step 4: Update the new membership matrix using the Equation, $\mu_i(x_i) = \frac{\left[\frac{1}{d_{ji}}\right]^{\frac{1}{m-1}}}{\sum_{k=1}^{C} \left[\frac{1}{d_{ki}}\right]^{\frac{1}{m-1}}}$, where m is a fuzzification parameter in the range [1.25,2].

Step 5: Go back to step2, unless the centroids are not changing.

An example of the Fuzzy C-means clustering is given in Fig. 1.21.

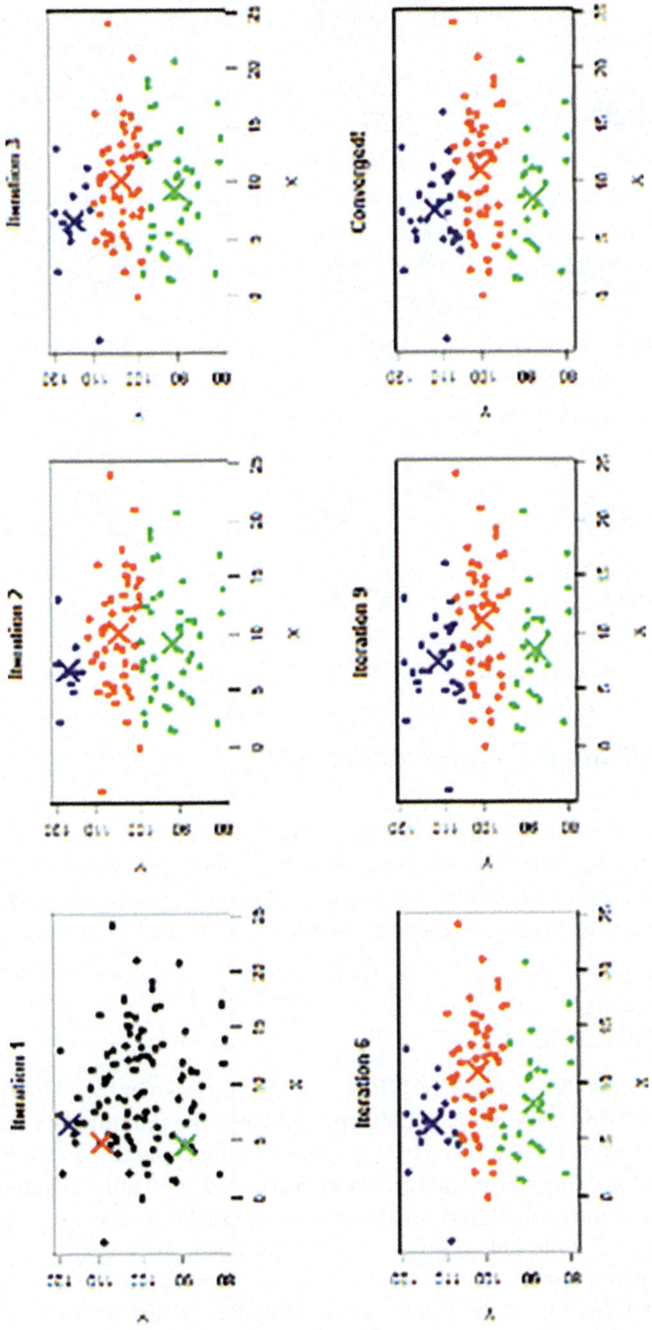

Fig. 1.20 K-means example

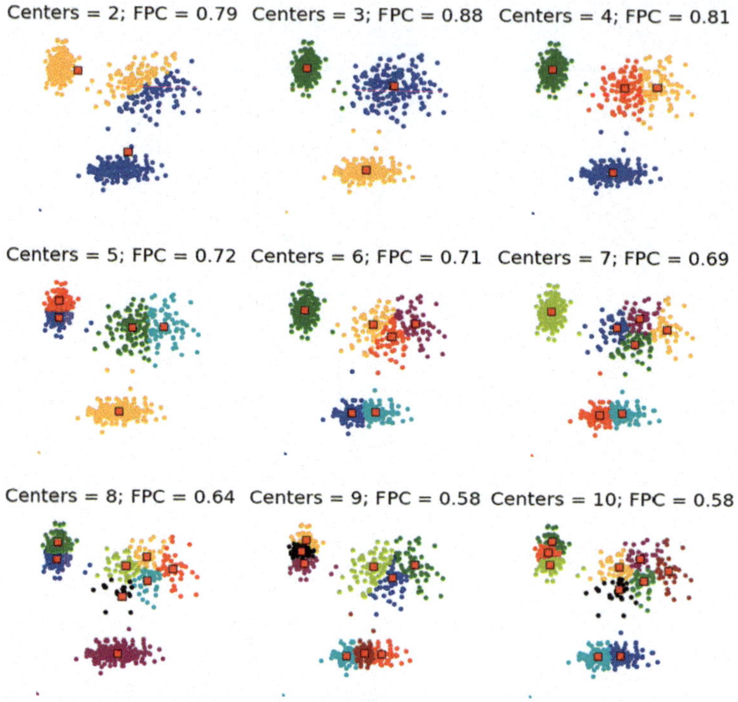

Fig. 1.21 Example of fuzzy C-means

1.1.9 Classification Techniques [25, 27]

Final step of any automatic prediction technique is classification. Classification deals with the labelling of test data based on training data containing observations whose labels are known, therefore supervised learning. Some of the pattern classification techniques, used in the book are described below:

(i) Naive Bayes
(ii) Neural Network
(iii) Gaussian Mixture Model

- Naive Bayes (NB) Classifier: NB is a statistical classifier based on Bayesian theorem. This classifier is mainly applied when the training data set should be adequate to represent the entire population that contains every combination of class label and attributes. It assumes that every feature is independent. It predicts the probabilities of the class membership that means the probability of a sample belonging to a particular class. To simplify the computation, it is assumed that the effect of an attribute value on a given class is independent of the values of the other attributes. This is called conditional independence. NB classifiers are very easy to implement and sometimes outperform over other sophisticated classifiers.

Consider a k class problem. To classify a new sample $X = (X_1, X_2, \ldots\ldots\ldots, X_l)$ where l is the feature dimension, the predicted output is defined in Eq. 1.15.

$$\hat{y} = \arg\max_{c_j \in C} P(c_j) \prod_{i=1}^{l} p(x_i|c_j) \tag{1.15}$$

where $\hat{y}$ denotes the class label predicted by the NB classifier and the possible class label is determined based on the highest probability of the class.

- Neural Network (NN): NN is the simulation of human nervous systems. The brain contains neurons, which act as organic switches. The network structure of the human brain is massively interconnected, where the output of a neuron can be forwarded to the thousands of neurons connected. NN [28, 29] can be viewed as a weighted directed graph in which artificial neurons are nodes and directed edges are connections between the neurons. A Multilayer perceptron (MLP) is a logistic regression classifier in which the input is first transformed using a learnt nonlinear transformation Φ and projects into a linearly separable space, referred to as a hidden layer. It consists of at least three layers of nodes. Except for the input nodes, each node is a neuron which uses a nonlinear activation function. MLP utilizes a supervised learning technique called backpropagation [30, 31] for training. Its multiple layers and non-linear activation function distinguish MLP from a linear perceptron. It can distinguish data that is not linearly separable. Figure 1.22 shows the architecture of an MLP with a single hidden layer.

A MLP with single hidden layer can be represented as $f : R^D \rightarrow R^L$ where D is the size of the input vector x and L is the size of the output vector $f(x)$, such that

$$f(x) = G\left(b^{(2)} + W^2\left(s\left(b^{(2)} + W^{(1)}x\right)\right)\right) \tag{1.16}$$

with bias vectors $b^{(1)}, b^{(2)}$; $W^{(1)}, W^{(2)}$ weight matrices and activation functions G and s.

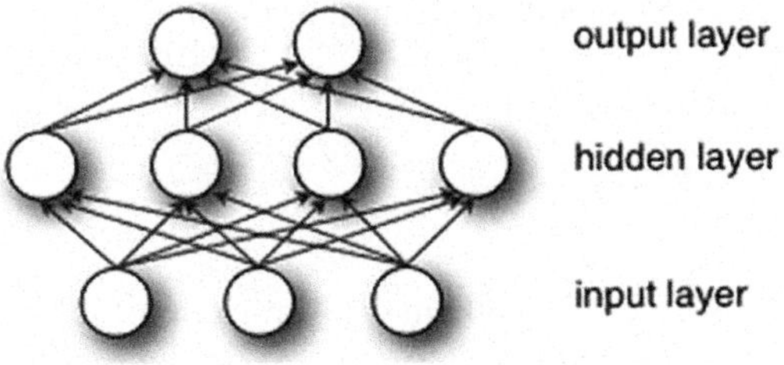

Fig. 1.22 MLP with a single hidden layer

The vector in the hidden layer is obtained using Eq. (1.17).

$$h(x) = \Phi(x) = s\left(b^{(1)} + W^{(1)}x\right) \tag{1.17}$$

Weight matrix $W^{(1)} \in R^{D \times D_h}$ connecting the input layer and the hidden layer neurons. The activation function (s) can be any one of the following

$$\tan \tan(a) = \frac{e^a - e^{-a}}{e^a + e^{-a}} \tag{1.18}$$

or using logistic sigmoid function

$$sigmoid(a) = 1/\left(1 + e^{-a}\right) \tag{1.19}$$

- Gaussian Mixture Model (GMM) [32]:

GMM is an unsupervised learning method used for soft classification of the data points based on probabilistic approach. Gaussian mixture is a parametric probability density function represented as a weighted sum of Gaussian component densities. The parameters are estimated from training data using iterative Expectation Maximization (EM) and the weighted sum of M component Gaussian densities is defined in Eq. (1.20).

$$p(\lambda) = \sum_{i=1}^{M} w_i g\left(X|\mu_i, \sum_i\right) \tag{1.20}$$

where X is a D-dimensional continuous vector, each w_i denotes the mixture of weights where, $i = 1, \ldots\ldots, M$ $g(X|\mu_i, \sum_i), i = 1, \ldots\ldots, M$ are the Gaussian densities, λ is used to represent parameter mean (μ) or standard deviation (σ).

The goal is to maximize the likelihood function with respect to the parameters, comprising the means (μ) and the covariance matrix ($\sum$) of the components and the mixing coefficients. GMM can be applied to a data set where the user knows that the data points are mixtures of a Gaussian distribution and form clusters with different mean and standard deviation. Figure 1.23 describes the functionality of GMM.

1.1.10 Petri Net

Petri Net is a mathematical modeling tool used to represent and analyze systems, characterized by concurrent, asynchronous, distributed, and non-deterministic behavior, i.e., originally developed by Carl Adam Petri in 1962. Petri nets have

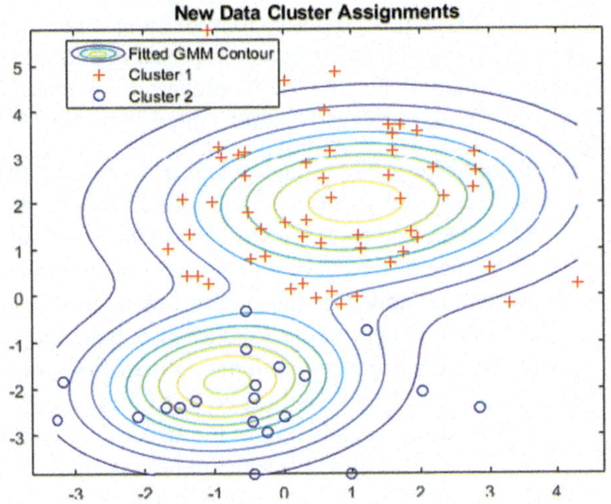

Fig. 1.23 GMM with two class problem

found applications in computer science, systems engineering, and workflow modeling. The key components of a Petri Net are:

(i) Places: These represent conditions, states, or resources in a system and are depicted as circles in graphical representations.
(ii) Transitions: It represents events or actions that can occur, changing the state of the system and depicted as rectangles or bars.
(iii) Arcs: These are used to connect places to transitions (and vice versa), showing the flow of tokens. Arcs can have weights, indicating the number of tokens required or produced.
(iv) Tokens: Represent the dynamic state of the system, which resides in places, and their distribution (marking) changes as transitions fire.

In a Petri Net, a transition fires when all its input places contain a sufficient number of tokens (as defined by the arc weights). Firing consumes tokens from input places and produces tokens in output places. This mechanism captures the dynamic behavior of the modeled system. It is a clear graphical representation of system dynamics and well-suited for modeling concurrency and synchronization. Figure 1.24 shows a representation of a Petri Net.

Here, the place node and the transition node are represented by a circle and a rectangular box, respectively. Directed arcs connect places and transitions, with

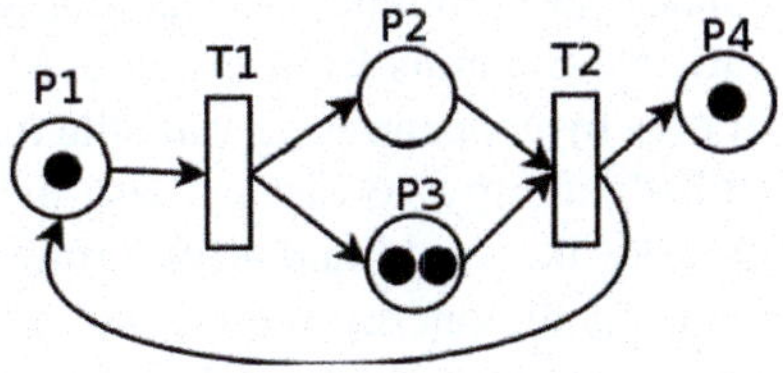

Fig. 1.24 A Petri net graph with four places and two transitions

some arcs directed from places to transitions and other arcs directed from transitions to places. The given example is composed of four places P1, P2, P3 and P4, and two transitions namely T1 and T2. There are many extensions of Petri Nets. Two major types are—Colored Petri Net and Timed Petri Net.

(i) Colored Petri Net—Each token of Colored Petri Net is attached with a color, indicating the identity of the token, which can distinguish the tokens. Also, each place and each transition has attached a set of colors. A transition can fire with respect to each of its colors, by which tokens are removed from the input places and added to the output places in the same way as that in traditional Petri Net, except that a functional dependency is specified between the color of the transition firing and the colors of the involved tokens.
(ii) Timed Petri Net—The Timed Petri Net consists of three specifications: topological structure, labeling of the structure, and the Firing rules. The topological structure of a Timed Petri Net generally takes the form that is used in a conventional Petri Net. The labeling of a Timed Petri Net consists of numerical values assigned to one or more of the components, like Transitions, Places, and Arcs, connecting the places and transitions. The firing rules are defined differently depending on the way the Petri net is labeled with time variables. The firing rules of a Timed Petri Net control the process of moving the tokens around. The Timed Petri Net can be classified into two major types:

    (a) Deterministic Timed Petri Net.
    (b) Stochastic Timed Petri Net.

(iii) Continuous Petri Net—This type of Petri Net has the same structure as Discrete Petri Nets, only the difference is that a transition in a continuous Petri Net fires continuously with a flow that may be externally generated by an input signal and may also depend on the continuous making.
(iv) Hybrid Petri Net: This is also an extension of traditional Petri Net with introduction of continuous places and continuous transitions or a mixture of discrete Petri Nets. So it has a discrete part and a continuous part. The discrete part models the logic functioning of a system and the continuous part models a system with continuous flow.

1.1.11 Problem Statement

Mineral processing, also known as ore beneficiation of mineral dressing, is a critical aspect of mining and metallurgy. It involves the separation and concentration of valuable minerals from raw ore. Mineral processing enhances the economic value of ores by concentrating valuable minerals and removing waste material. In Mining and Metallurgy, to extract base metals (like copper, zinc, and lead) and precious metals (like., gold and silver), mineral processing is required. In the construction industry, the energy sector, the chemical industry, aerospace, and defence everywhere, the use of mineral processing can be found. Nowadays, it has been seen that the need for mineral processing is increasingly gaining importance. One of the important areas of mineral processing is grinding. The existing grinding processes are, however, not able to predict the final product and therefore, do not provide a legitimate solution. Although a certain quality restriction is always imposed on the production of minerals, economics ultimately dictate the market potential of the mined-out product. On this basis, at least the following basic considerations on the process must be adjudicated in the work:

(i) To control the particle size grinding process,
(ii) To generate a product of regular size and its composition,
(iii) To expose or liberate constituents for subsequent processing,
(iv) To Control the Composition.

Mineral products inherently lack uniformity in size distribution, their valuation is thus a complex issue involving credits for desirable features, and simultaneously penalties for undesirable features. Upgrading the quality of minerals, and its adequate liberation for subsequent uses broadly is the scope of mineral processing [5]. In summary, the processing of minerals can broadly be classified into four categories:

(i) Comminution
(ii) Process separation
(iii) Medium recovery and
(iv) Product recovery.

Though comminution or crushing is not directly involved in separating minerals from the waste, it is crucial in liberating undesirable products in the entire operation. Any crushing operation generally includes an input leading to the final product and waste component. One point at this junction is absolutely clear, the grade (quality) and recovery represent a trade-off for a given sample of materials and are solved as a multi-objective optimisation problem.

An important aspect of any mineral processing operation is material balance. Theoretically, no material is destroyed despite their size-mix change during crushing. However, a mathematical relationship that represents the breakage characteristics during the crushing process is not obtained yet. More particularly, the invention relates to the development of an intelligent expert control device enabling optimum grinding by a processing ball mill to output ground products with desired particle size distribution. To perform this, the main concern is to minimise the wastage of raw material by controlling the ball mill operation to achieve desired particle size ranges. The experiments are conducted in a noise-free environment, and it has been noticed that the acoustics of the ball mill vary gradually with crushing. The change is significant, and one can easily recognize that the particles are grinded, generating larger to smaller particles. Based on this observation, it has been decided to install microphones outside the ball mill for capturing signals during grinding. The acquired signal is pre-processed and analysed using computational methods to learn the acoustic signatures at different stages of grinding with various raw materials. The acoustic signatures are stored along with the corresponding particle size distribution and used to predict the desired particle size distribution for a new feed size and input mix.

1.1.12 Innovation and Research: Main Contribution of This Volume

The remaining Chapters of this Handbook are as follows:

- Chapter 2: This Chapter is dedicated to the acoustics of the Ball Mill sound. Classify different states of the Ball Mill, in its load-varying condition and empty condition. Techniques are discussed to prepare a data set for real-time analysis of the mill acoustics and based on the data set the different running states are derived. Finally, manual state identification is compared with Experimental states.
- Chapter 3: This Chapter has mainly focused on particle breakage modelling. This Chapter aims to describe the development stages of an appropriate control schematic for dynamic breakage concerning particle size distribution.
- Chapter 4: This Chapter discusses the acoustics-based modelling of the ball mill sound and the technique discussed to identify different load-varying conditions. A mathematical model is derived from the states based on acoustics parameters and ultimately the results are discussed.
- Chapter 5:This Chapter is dedicated to finding the correlation of derived states with the particle size distribution. An evolutionary algorithm is used to predict the next from the current state. Then the particle size distribution predicts to reduce the wastage of materials.
- Chapter 6: This Chapter discusses a research background to design a real-time Intelligent system. Using the method discussed, the machine can be stopped automatically whenever the desired particle size distribution is achieved.
- Chapter 7: This Chapter demonstrates how Petri Net has been used successfully for implementing, monitoring, and controlling the activity of a tumbling mill to achieve optimum grinding within residence time.
- Chapter 8:This Chapter concludes the book and provides the future research path.

References

1. Gills, W.A.: Mineral Processing. Springer Verlag, Germany (1991)
2. Taggert, H.: Hand Book of Mineral Processing. SME Publishers (1951)
3. Vesilind, P.A.: The Rosin-Rammler particle size distribution. Resour. Recover. Conserv.. Recover. Conserv. **5**(3), 275–277 (1980)
4. Vitez, T., Travnicek, P.: Particle size distribution of a waste sand from a waste water treatment plant with use of Rosin-Rammler and Gates-Gaudin-Schumann mathematical model. Acta Universitatis Agriculturae et Silviculturae Mendelianae Brunensis **59**(3), 197–202 (2014). https://doi.org/10.11118/actaun201159030197
5. Wills, B.A.: Mineral Processing Technology. Pergamon Press, UK (1986)
6. Monov, V., Sokolov, B., Stoenchev, S.: Grinding in ball mills: modeling and process control. Cybern. Inf. Technol. **12**(2), 51–68 (2012). https://doi.org/10.2478/cait-2012-0012
7. Lameck, N.N.S.: Effects of grinding media shapes on ball mill performance. Thesis paper, Johannesburg, October 2005

8. Fuerstenau, D.W., Kapur, P.C., De, A.: Modelling breakage kinetics in various dry comminution systems. KONA Powder Part. J. **21**, 121–132 (2003)
9. Campbell, J., Spencer, S., Sutherland, D., Rowlands, T., Weller, K., Cleary, P., Hinde, A.: SAG mill monitoring using surface vibrations. In: Proceedings International Autogenous and Semiautogenous Grinding Technology, pp. 373–385 (2001)
10. Henderson, P.D.: The fundamentals of FFT-based ® audio measurements in smart live. Notes, pp. 1–11 (2020)
11. Heckbert, P.: Fourier transforms and the fast Fourier transform (FFT) algorithm. Comput. Graph.. Graph. **2**, 15–463 (1995)
12. Harčarik, T., Bocko, J., Masláková, K.: Frequency analysis of acoustic signal using the fast Fourier transformation in MATLAB. Procedia Eng. **48**, 199–204 (2012). https://doi.org/10.1016/j.proeng.2012.09.505
13. Prahallad, K.: Speech Technology: A Practical Introduction Topic: Spectrogram, Cepstrum and Mel-Frequency Analysis. Carnegie Mellon University & International Institute of Information Technology Hyderabad (2011)
14. Rao, K.S., Manjunath, K.E.: 'MFCC Features' 85, Speech Recognition Using Articulatory and Excitation Source Features, pp. 81–88. Springer Briefs in Speech Technology (2017). https://doi.org/10.1007/978-3-319-49220-9
15. Castells, F., Laguna, P., Sörnmo, L., Bollmann, A., Roig, J.M.: Principal component analysis in ECG signal processing. EURASIP J. Adv. Signal Process. **2007**(74580), 1–21 (2007). https://doi.org/10.1155/2007/74580
16. Bhattacharjya, R.K.: Introduction to Genetic Algorithms. Department of Civil Engineering IIT Guwahati, Notes (2013)
17. Beasley, D., Bull, D.R., Martin, R.R.: An overview of genetic algorithms: part 1, fundamentals. Univ. Comput. **15**(2), 56–69 (1993)
18. Murthy, C.A.: Genetic Algorithms: Basic Principles and Applications. Machine Intelligence unit Indian Statistical Institute (2013)
19. Walck, C.: Hand-book on statistical distributions for experimentalists. Internal report SUF–PFY/96–01. Stockholm, pp. 1–202 (2007)
20. Forbes, C., Evans, M., Hastings, N., Peacock, B.: Statistical Distributions, 4th edn, pp. 1–50. Willy Publications (2011)
21. Lahodny Jr., G.: Curve Fitting and Parameter Estimation, pp. 1–18 (2015)
22. Zucchini, W.: Applied Smoothing Techniques Part 1: Kernel Density Estimation, pp. 1–20 (2003)
23. Gurley: Curve fitting techniques. In: Numerical methods lecture 5, CGN 3421—computer methods, pp. 86–99 (2018)
24. Hochreiter, S.: Machine Learning: Supervised Techniques. Lecture Notes, pp. 119–159 (2014)
25. Smola, A., Vishwanathan, S.V.N.: Introduction to Machine Learning. Cambridge University Press, pp. 20–36 (2008)
26. Cannon, R.L., Dave, J.V., Bezdek, J.C.: Efficient implementation of the fuzzy c-means clustering algorithms. IEEE Trans. Pattern Anal. Mach. Intell.Intell. **8**(2), 248–255 (1986)
27. Jacob, E.K.: Classification and categorization: a difference that makes a difference. Libr. Trends. Trends **52**, 516–551 (2004)
28. Abiodun, O.I., Jantan, A., Omolara, A.E., Dada, K.V., Mohamed, N.A., Arshad, H.: State-of-the-art in artificial neural network applications: a survey. Heliyon **4**(11), e00938 (2018). https://doi.org/10.1016/j.heliyon.2018.e00938
29. Gao, S., Zhou, M., Wang, Y., Cheng, J., Yachi, H., Wang, J.: Dendritic neuron model with effective learning algorithms for classification, approximation, and prediction. IEEE Trans. Neural Netw. Learn. Syst. **30**(2), 601–614 (2018)
30. Reynaldi, A., Lukas, S., Margaretha, H.: Backpropagation and Levenberg-Marquardt algorithm for training finite element neural network. In: 2012 Sixth UKSim/AMSS European Symposium on Computer Modeling and Simulation, pp. 89–94. IEEE (2012). https://doi.org/10.1109/EMS.2012.56

31. Rumelhart, D.E., Hinton, G.E., Williams, R.J.: Learning representations by back-propagating errors. Nature **323**, 523–526 (1986)
32. Anand, V., Pushp, D., Raj, R., Das, K.: Gaussian mixture model (GMM) based object detection and tracking using dynamic patch estimation. In: 2019 IEEE/RSJ International Conference on Intelligent Robots and Systems (IROS), pp. 4474–4481. IEEE (2019). https://doi.org/10.1109/IROS40897.2019.8968275

Chapter 2
Ball Mill Acoustics and Analysis

Abstract This chapter is dedicated to the acoustics of the running ball mill sound which has been analysed in order to predict the empty and loaded condition of the running ball mill. Here, the understanding of the working ball mill based on acoustic has been discussed that is used to interpret the characteristic of particle breakage phenomena. During the grinding operation, the acoustics of the running ball mill loaded with raw materials having different particle size distribution and empty conditions are captured using sensors. The continuous acoustics produce voluminous data which is difficult to analyse manually in real-time scenarios. Data preparation techniques for real time analysis of data to derive different states of running conditions of the ball mill are discussed in this chapter. Simulation results obtained through empirical modelling of the acoustic signals are compared with experimental states in order to fine tune the parameters of the model. Finally, the model is designed to identify different running status (empty or loaded) of the ball mill.

2.1 Introduction

A ball mill [1, 2] is a grinding equipment widely used across various industries. Its primary purpose is to grind and blend materials into fine powders or smaller particles through the impact of grinding media against the raw material within a rotating drum. This grinding process enables finer particle sizes, which is essential for processes like mineral liberation.

Ball mill sound analysis plays a crucial role in optimising operational efficiency, minimising energy consumption, and ensuring product quality. The sound generated by a ball mill is indicative of the grinding process when analyzed and produces knowledge about the operating conditions of the mill. The noise produced by the mill can reveal whether it is under or over-filled with the material or whether the grinding media (e.g., balls) are making proper contact with the material. This real-time feedback helps the operators to maintain optimal load levels for efficient milling.

During the operation of the mill, the human ear can easily notice the change of sound with different loading or empty conditions of the ball mill. Sound analysis

J. Sil et al., *Intelligent Real-Time Control Systems to Predict the Particle Size of Minerals Using Acoustics of Ball Mill*, Studies in Systems, Decision and Control 600,
https://doi.org/10.1007/978-3-031-91290-0_2

helps to determine the impact of grinding media on the material. By adjusting the mill parameters based on sound frequency and amplitude, operators can achieve a finer grind or reduce unnecessary wear on grinding media and mill liners. Variations in sound indicate changes in the material grind size, allowing operators to fine-tune the milling process for better control over product particle size and uniformity. Monitoring the running ball mill's sound can help the operators to detect inefficient operating conditions, such as excessive load or lack of material in the ball mill, both of which increase power consumption. Properly adjusted load levels reduce the energy needed per ton of material processed, minimising operational costs. Sound analysis can be integrated into automated control systems to maintain these optimal conditions, thereby saving energy consistently over time. The analysis can also indicate issues with internal components, such as worn liners, broken grinding media, or improper mill speed. Changes in mill sound can help to detect these issues early, allowing for proactive maintenance and minimising unplanned downtime. Predictive maintenance based on sound patterns improves equipment lifespan and helps to prevent catastrophic failures, saving repair costs and maintaining production schedules.

Achieving a consistent particle size is essential for industries where precise product specifications are required. By hearing, if analysing the sound profile is possible, then operators can quickly detect deviations in the grind size and adjust accordingly to produce a uniform product. Consistent quality not only meets regulatory standards but also enhances customer satisfaction. Variations in sound can reveal potentially hazardous conditions, such as overloading or unbalanced grinding media, which can lead to structural failures in the mill. Early sound warnings can allow operators to take corrective actions, and reduce the risk of accidents or damage to equipment and personnel. Sound analysis can be part of an integrated control system, using advanced sensors and artificial intelligence to maintain optimal mill conditions autonomously. With automated sound-based control, the mill can dynamically adjust its operational parameters (e.g., speed, load, water addition) without human intervention, maximizing efficiency, reliability and, most importantly, wastage of the mineral.

Analysis of sound enhances process control, reduces operational costs, ensures product quality, and supports predictive maintenance. By leveraging sound data, operators can improve overall mill performance, contribute to sustainability efforts, and create a safer working environment. The aim of the research work, described in this chapter, is to identify different working conditions of the ball mill by analysing the sound of grinding operations. A framework has been proposed to identify the working condition of a ball mill by analysing the data. The method distinguishes the loaded mill from an empty mill. Using this concept, the mill is controlled in its closed-loop environment at run time. The model is valid for other different kinds of reactors/crushers/compound machines.

2.2 Components of a Ball Mill

A typical ball mill consists of several key components:

- Cylinder or Drum: The main chamber/cylinder/vessel, often made of steel or a lined metal, where grinding takes place. It can vary in size depending on the scale of production, from small laboratory mills to large industrial mills several meters in diameter.
- Grinding Media: The balls or rods inside the drum that grind and pulverise the material. These grinding media can vary in size and material, depending on the type of mill and the specific grinding needs.
- Feed and Discharge Mechanism: The ball mill includes an inlet where raw materials are fed into the cylinder and an outlet where the processed material exits. In some mills, the material may be fed continuously, while others operate on a batch basis.
- Drive Mechanism: The system that powers the rotation of the drum, typically including a motor, gears, and sometimes a pulley system, depending on the design. The speed of rotation can influence the effectiveness of the grinding process.
- Liners: The interior of the drum is often lined [3] with replaceable materials such as rubber or manganese steel. These liners protect the drum and influence grinding efficiency by modifying the motion of the balls.

2.2.1 Operation of a Ball Mill

During the operation of the ball mill [4], the rotating cylinder causes the grinding media to lift and then cascade or tumble back down. The impact force created by this tumbling motion, combined with the sliding and rolling of the grinding media against the material, leads to the pulverization of the input feed into finer particles. The ball mill's rotational speed, the size of the grinding media, and the load of the material in the drum all contribute to the effectiveness of the grinding process.

There are two primary types [5] of ball mills based on the way they handle the material:

- Wet Ball Mill: In this type, water or another liquid is added to the drum to form a slurry, which helps to control dust and improve efficiency.
- Dry Ball Mill [6]: This type operates without any liquid, relying solely on the dry grinding process.

2.2.2 Types of Motion Inside Ball Mill

The motion within a ball mill can generally be divided into three types, based on the speed and load [1] conditions:

- Cascading: At lower speeds, the grinding media cascade against the wall of the drum and produce a tumbling motion.
- Cataracting: As speed increases, the grinding media begins to fall freely, creating a more impactful grinding action.
- Centrifuging: At very high speed, the centrifugal force pushes the media against the drum walls, and effective grinding decreases as there is less tumbling action.

2.3 Mechanics of Sound Generation in a Ball Mill

First, we find the primary sources of the sound [7, 8] within a ball mill. There are two sources of sound production [9] within a running ball mill. The following are the sources.

2.3.1 Grinding Media

The ball mill rotates around a horizontal axis, causing the grinding media to be lifted on the rising side of the mill due to centrifugal forces. As the balls reach a certain height, they fall back down due to gravity. This falling motion creates a cascading effect, where ball's impact each other and the material, causing size reduction through impact. At higher rotation speeds, the grinding media can be thrown further up along the ball mill wall, creating a cataracting motion. This means that the balls are lifted to a greater height before they fall onto the material below, which increases the impact force. When the grinding media collides with the material in this impact zone, it fractures and reduces particle size effectively. As the grinding media falls onto the material, compression forces are applied, crushing it between the balls and the mill lining.

Additionally, grinding media in contact with each other and the mill lining create frictional forces. This produces attrition, a shear force that further reduces particle size, especially for finer grinding. Here, the design of the mill liners has an important usage. Liners with lifters or ridges help to lift the media higher, enhancing the cascading and cataracting action, while smooth liners encourage a more gentle tumbling action. As the media tumbles over one another and the liners, material is caught in the spaces between them, causing more collisions and increasing grinding efficiency. The generated sound also depends on the ball mill's rotational speed, which is critical. Too slow, and the balls will merely roll over each other, causing less impact force. Too fast, and the balls will be pinned to the wall by centrifugal force, preventing effective impact. Optimal speed, often around 65–85% of the critical speed, ensures a balance where cascading and cataracting are maximized, enhancing collision frequency and grinding efficiency. Different materials generate distinct sound profiles based on their density, size, and hardness. For instance, harder materials produce sharper, higher-pitched impact sounds, while softer or finer materials create more muted, lower-frequency noises. This characteristic sound profile helps operators identify changes in material properties or loading conditions.

2.3.2 *Grinding Media and the Ball Mill Shell*

The sound generated due to the contact between the grinding media and the mill shell or liners is considered as noise. When grinding media balls fall, they collide with both the material being ground and the mill liners. This collision generates impact forces, which create vibrations and sound waves that are transmitted through the mill shell, producing a characteristic grinding noise. The higher the speed and the larger the size of the balls, stronger impact has been observed and consequently louder the generated noise. The material being processed also collides with the liners as it moves inside the ball mill. These collisions create a scraping or grinding noise, depending on the size, density, and hardness of the material. If the ball mill is underloaded, the grinding media have more space to collide directly with the shell and liners, resulting in louder impact sounds. Conversely, if the ball mill is overfilled, the material dampens the sound by cushioning the impact between the grinding media and the liners.

The design and shape of the ball mill liners significantly influence the noise. Liners with ridges or lifters are designed to lift the grinding media higher before it falls, increasing the impact force and, consequently, the noise level. More aggressive lifter profiles (i.e., higher ridges) cause the media to fall from a greater height, producing louder sounds upon impact. Over time, as liners wear down, the impact characteristics change, often resulting in variations in noise frequency and intensity. Apart from direct impacts, a portion of the noise is generated by the tumbling motion of grinding media against the liner. As the grinding media and material slide and roll along the liners, they produce a rumbling or scratching noise. This frictional sound is more pronounced when the ball mill is operating with a larger volume of material, as the interaction between the balls and the material increases. The noise produced by the ball mill shell and liners varies with the mill's operational conditions, such as load levels, rotational speed, and liner wear. A higher speed generally leads to higher frequency noise, while a ball mill with a well-maintained load level and optimal speed produces a more rhythmic, continuous sound. Deviations from this typical sound profile often indicate the operational issues.

2.3.3 *Factors Affecting Sound Characteristics*

The sound of a ball mill is an indicator of the speed and the grinding efficiency. Optimal grinding occurs when the speed produces a balanced and rhythmic impact sound, while speeds that are too low or too high can lead to inefficient grinding and specific sound patterns. There are several factors affecting the sound characteristics

[10, 11] of the ball mill. Those are operating speed, load levels, linear and media material and ball mill condition and wear [3, 12]. The following are the characteristics [13] in detail:

- Operating Speed

The ball mill can operate at low, high and moderate speeds [14]. At low speed, the balls inside the vessel slide or cascade over each other rather than rising high and falling. This produces a soft, muffled sound due to lower impact forces between the balls and the ball mill liner. The grinding action is primarily abrasive, and the sound is quieter and more even. As the speed increases to an optimal range, the balls are lifted higher, creating a balance between cascading and cataracting motions (where balls start to tumble). This speed range leads to increased impacts as the balls fall from greater heights, resulting in a louder, rhythmic sound. This is the most effective range for grinding efficiency, with a balanced, steady sound indicating good grinding action. Finally, when the ball mill operates near or above the critical speed, the centrifugal force becomes strong enough that the balls are pressed against the ball mill wall, reducing cascading and cataracting. This causes the grinding media to stick to the ball mill wall, reducing impact collisions. The result is a quiet, almost silent operation due to the lack of impacts. This condition is generally undesirable for grinding efficiency. However, if the speed increases thereafter, going significantly beyond the critical speed, the ball mill can produce a very irregular sound or even "mill noise," indicating mechanical issues due to excessive vibrations or erratic movement of grinding media. This condition may damage the ball mill and reduce grinding efficiency.

- Load Level of the Ball Mill

The sound of the ball mill depends on the load level of the materials present within the ball mill. In the underloaded ball mill, too few balls [15] or materials are present, so there are frequent impacts between the balls and the mill shell. This produces a loud, sharp, and more metallic sound due to the direct contact of the balls with the shell. The lack of cushioning from sufficient material also means the sound is uneven and fluctuates as the balls hit the shell frequently. Grinding efficiency is often low in this case because there isn't enough material to absorb and optimize the impact energy. At an optimal load level, where there's an ideal balance of balls and material, the sound of the mill is steady, rhythmic, and moderate in volume. This condition generally indicates efficient grinding and is often the desired operational state. If the mill is overloaded, then there is too much material in comparison to the balls. This reduces the amount of cascading action, and the balls are not lifted as high before falling. The sound, in this case, is muted, dull, and low in volume because the balls are cushioned by the excessive material. Grinding efficiency is typically reduced, as the energy transfer between balls and material is less optimal due to the overcrowded conditions. In some extreme cases, where the mill is significantly overloaded, the sound can become almost inaudible as the grinding media movement is restricted. This "silent" mill condition indicates poor grinding efficiency and potential mechanical strain. The sound of a ball mill becomes louder and sharper with lower loads due to increased impacts on the mill shell, while a balanced, steady sound typically indicates an optimal load with effective grinding.

- Liner and Media Material

The liner material impacts the sound produced during the operation of the mill. The mill may have a hardliner or soft liner. Hard-liners tend to produce a sharper, louder sound when the grinding media impacts them. These materials don't absorb much of the impact energy, so they result in more pronounced, metallic sounds due to the high force of collisions between the media and the liners. Steel liners, for example, will generate a high-pitched, ringing sound due to their rigidity. Liners made of rubber or softer materials treated as soft liners produce a more muted, lower-pitched sound. These materials tend to absorb more of the impact energy, reducing the sharpness of the collisions. The sound produced is often less harsh and can be described as more "dull" or "thumpy." Rubber liners also help to reduce wear on both the media and the mill shell, which can lead to a quieter operation in some cases.

- Grinding Media Material

The grinding media (balls or rods) inside the mill also plays a crucial role in the sound produced. The material of the media affects the way it impacts both the liner and other media, influencing the sound in different ways. The size and shape of the grinding media also affect the sound. Larger balls tend to produce a deeper, more resonant sound because they hit the material with more force, whereas smaller balls may create a higher-pitched, more frequent sound due to quicker impacts. Irregular shapes (e.g., cylpebs or ellipsoids) can create different sound profiles compared to spherical media, typically producing more varied or irregular noise due to their less predictable motion. The media is made of steel, chrome, ceramic, lead or rubber-coated. Steel and chrome balls considered as the hard grinding media produce louder, sharper sounds because they create more powerful impacts when they collide with the material or liner. These media generate a distinctive "clanking" or "ringing" sound, especially in dry milling operations. The higher the density and hardness, the more pronounced the sound.

However, Softer media materials tend to create a quieter, more subdued sound because they absorb more of the impact energy. Ceramic balls, for instance, can produce a lower-pitched sound that is less harsh compared to steel balls. This softer sound is typically indicative of a less energetic grinding action, which may be beneficial for delicate materials or processes that require lower-intensity grinding.

- Raw Materials

In different industries, the ball mill is used for grinding various materials. Copper, Gold, Iron, Silver Zinc Ores, Phosphate Rock, coal, coke, limestone, etc., are used in the mining industry, fertilisers and other agricultural products, energy production and steel manufacturing cement production, agriculture, and as industrial fillers respectively. The mill with different types of material also produces different sound patterns

[16]. Each material requires specific milling parameters (like speed, grinding media type, and mill design) to achieve the desired particle size, texture, and quality. For different material types, hardness, size and amount affect the generated sound for the mill.

2.4 Sound Produced During Ball Mill Operation

The ball mill has a vessel which contains the raw material to be grind. The mill can rotate on its axis in its running condition. A mill may be empty if it runs only with the grinding media. On the other hand, the mill may run in different load-varying conditions with grinding media and raw materials. The produced sound with the same number and sized grinding media is different for different raw materials. In the same way, a particular grinding material with the same number and sized grinding media produces different sounds. So, the sound changes with varying types of raw material, weight of the raw material, grinding media type and media size. To experiment using sound, among those four parameters—raw material type, weight, grinding media number and size, three parameters need to be constant. The running ball mill sound is captured in a noise-free environment. Since the mill produced a loud sound in its operational condition, all possibilities of the outside noise needed to be nullified. In the loaded condition, the mill was allowed to run continuously to crush the material and the produced sound during the operation is captured using the arrangement of the microphones, discussed in Chap. 3.

Figure 2.1 shows the time-domain acoustic representation of the captured sound of a mill during its load-varying condition. The sound of the empty mill with only the grinding media is also captured, which is shown in Fig. 2.2. In both the load-varying and empty conditions, the number and the size of the grinding media considered are the same. In load-varying conditions, a particular type of raw material is considered for the whole operation till the material reaches its fine production. In the empty condition, the repetition of the signal pattern occurs after a few seconds, as observed while analysing the sound spectrum manually, therefore, recording beyond one minute does not provide any meaningful features. But in load-varying conditions, the sound is captured till it's fine production.

2.5 Transformation of the Sound Signal

The raw data of the recorded sound depends on the type of microphone or the sensor used during the sound recording of the running mill. If a monophonic microphone is used for audio capture, then the raw data have a vector as the captured sound since the microphone produces a single channel. However, if a stereophonic microphone is used, then the raw data is produced as a matrix of two columns since the microphone

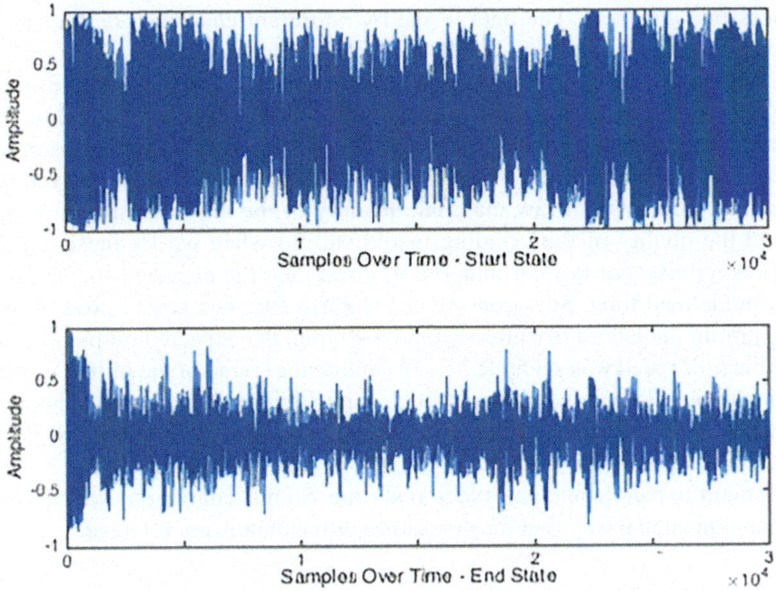

Fig. 2.1 Acoustics of the loaded ball mill

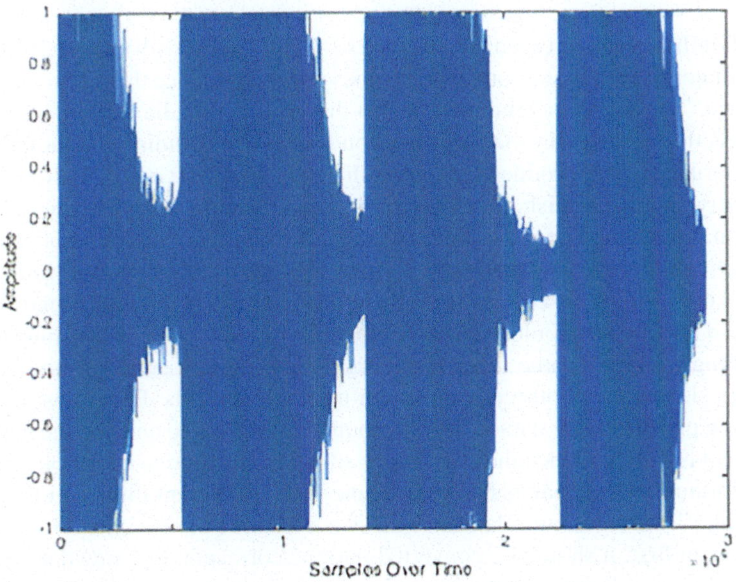

Fig. 2.2 Acoustics of an empty ball mill

produces two channels. The data in the two different channels are the same. The recorded acoustic can be stored in "WAV" file format.

The stored raw data is a time domain continuous data. It represents the amplitude of the sound with respect to zero. Generally, for a small segment of sound, say, for one second of capturing sound, around 10 K raw data (approximately) is produced. To generate an expected fine of the material, different factors are involved such as type of ball mill, type of raw material, quantity of the material, speed of the mill, size and the quality of the grinding media etc. So when we fix these parameters and allow a particular type of material to grind then the expected fine is produced within predefined time. Say, iron ore of weight 2.7 kg is taken to produce the fine of 220 μm. In the laboratory after around 9.45 min, the expected output is produced where the mill speed was set 50 R.P.S. To capture the sound of the mill in its runtime, a monophonic microphone is used, which captures the data using one channel. The continuous sound is now converted into the time domain producing raw data. The size of the produced data captured in the laboratory is more than 10,000 K, which is very difficult to handle and analyse in real time. So this continuous data needs to be segmented in such a way that the continuity of the data is not affected.

2.6 Different States of Operation

The ball mill can run in two different states—Empty and Load-varying. If an empty vessel rotates, it produces a sound which does not change since there is no operational difference. The sound is generated due to the grinding media falling on the vessel liner directly, whereas, in load-varying conditions, the grinding media falls on the raw material. So, the produced sound is different from the empty ball mill sound. To identify and distinguish the empty mill sound from the sound of the loaded mill condition, both sounds need to be analysed. For this, the mill sound is captured in its empty condition as well as load-varying condition. The empty ball mill sound is captured only for a few seconds since there is no change in the mill sound. Whereas the loaded mill sound should be captured in terms of minutes since finishing the work of crushing may take some time. Handling the empty mill sound is not a problem since the size of the captured sound is in terms of seconds. For a few seconds of sound, the produced raw data is of size around 100 K. But to analyse the loaded ball mill sound is difficult since the continuous sound may generate enormous data. This large continuous data analysis and extracting some important information from it is a challenge. This amount of data analysis is not feasible in run time. After capturing the sound, at first, it should be converted into the corresponding raw data. Since the data is in the time domain, so it will produce the amplitude with respect to zero as the raw data. Figures 2.1 and 2.2 both show the time domain representation. Depending on the experimental outcome, the following steps are considered to obtain the data suitable for analysis:

(i) A ball mill vessel rotates on its axis vertically. Ball mill rotates differently based on its motor speed and size. During the manual analysis of the signal, it is noticed that the ball mill rotates a full cycle on its axis after each fixed amount of time. To keep the continuity of the sound signal and easy analysis of the captured sound, the continuous signal is divided into fragments based on one full cycle of rotation of the ball mill. Therefore, to analyse the data, the segment is based on one full rotation of the mill vessel. After each full cycle of rotation, the acquired signal has been segmented where each segment is approximately equal in size. Each crushing operation with different raw materials and with different weights takes a different amount of time to produce the desired particle size. Depending on the type of raw material and the total time taken by a specific mill to produce the desired particle size, the stages of operation can be decided. Here, an example is considered where iron ore is taken as the raw material and for a laboratory-based mill, it takes 8 min to complete the grinding operation as per the requirement. Four different segments at four different stages of crushing with the same input mix are shown in Fig. 2.3. The crushing starts with the input mix, which contains raw material of large size, approximately 6–10 mm and is grinded by the grinding media made of iron. At the beginning of crushing, the impact and attrition between the material and the grinding media are very high. In this period, it generates a huge amount of sound, and the amplitude of the samples is very high, which is shown in Fig. 2.3a. But gradually, as the crushing operation is going on towards its end, the impact becomes faint, and the size of the material has been reduced, resulting in amplitude representation in a smaller range. Figure 2.3b and c show the sound generated during the internal stages of grinding. At the end, since the crushing is almost complete, the impact between the material and the media is low, as depicted in Fig. 2.3d.

(ii) During the operation of the mill, the human ear can easily notice the change of sound when the ball mill traverses from one stage of crushing to another stage. After the completion of each cycle, the mill moves on to the next cycle and the fraction of time of mill operation is considered as the transition period of the mill acoustics. For example, the amplitude variation of a laboratory ball mill is in the range (– 0.8) to (+ 0.8) during the transition period when coal is used as crushing matcrial. Based on this observation, the signal is segmented, and the corresponding amplitudes are stored in arrays. The plots of the transition data set in different stages of load-varying conditions are shown in Fig. 2.4 with a range of amplitude variation at different states. At the first state it is approximately $+ 0.8$ to $- 0.8$, which gradually reduces and reaches approximately $+ 0.35$ to $- 0.35$ at the end of crushing. The amplitude band for each state is highlighted in the Fig. 2.4a, b, c and d.

(iii) The same segmentation procedure can be applied to the captured sound of the empty ball mill. Figure 2.5a and b show one full cycle and one transition cycle of the empty ball mill, respectively. The empty ball mill segment structure is

(a)

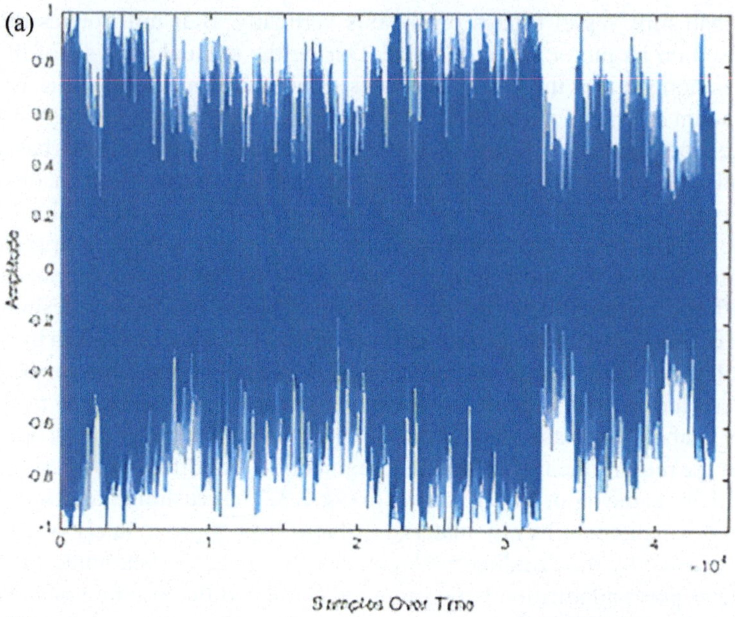

(b)

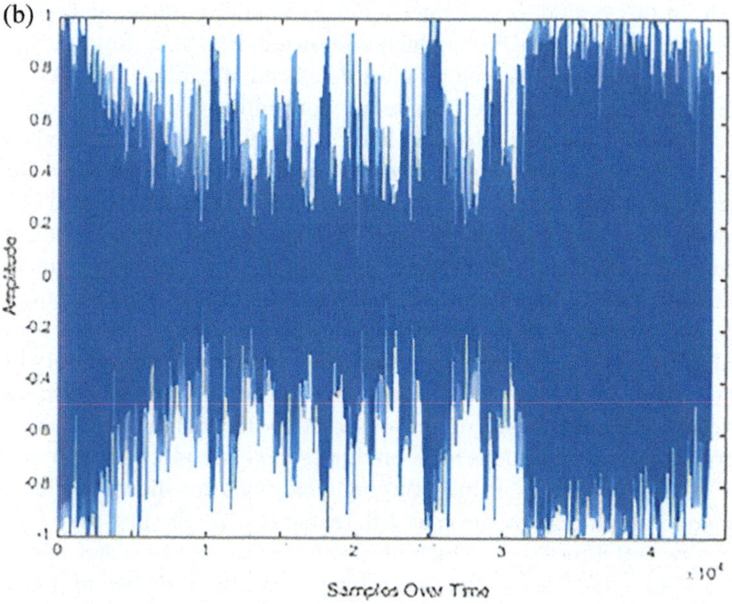

Fig. 2.3 **a** At the beginning of crushing. **b** One minute after crushing starts. **c** One minute before the end of crushing. **d** End of crushing

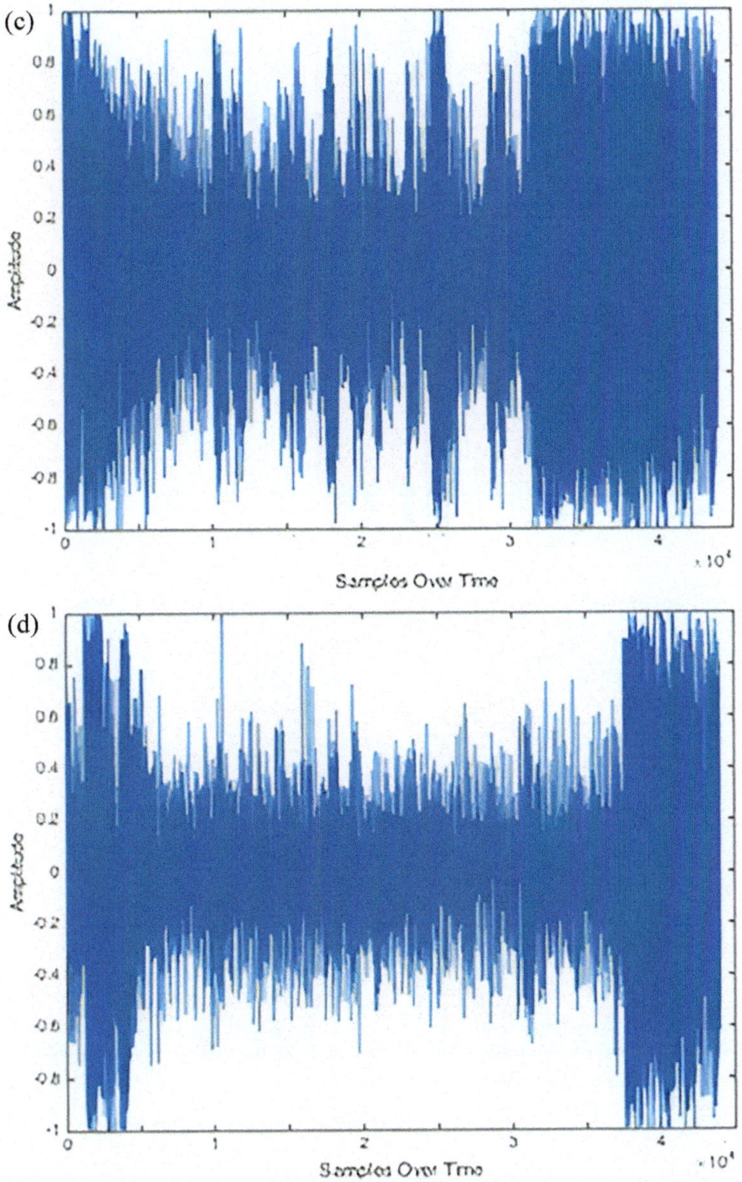

Fig. 2.3 (continued)

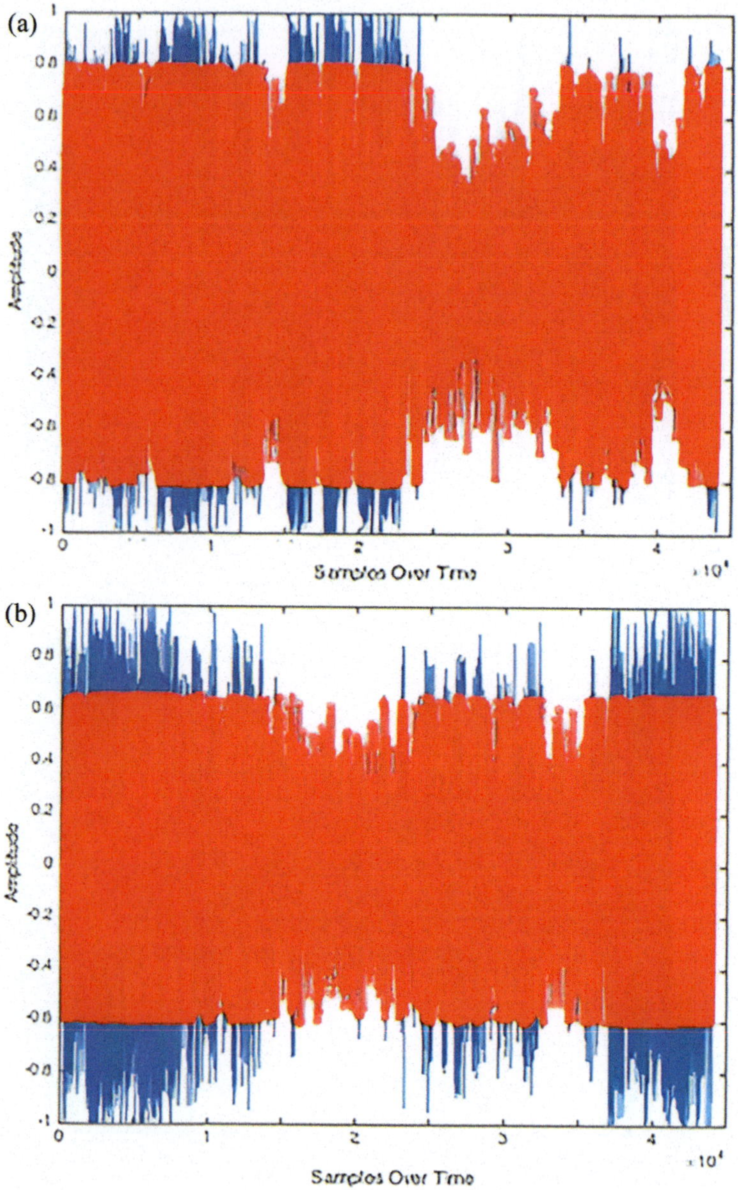

Fig. 2.4 **a** Transition segment at the beginning of crushing. **b** Transition segment after two minutes of crushing. **c** Transition segment before two minutes of end of crushing. **d** Transition segment at the end of crushing

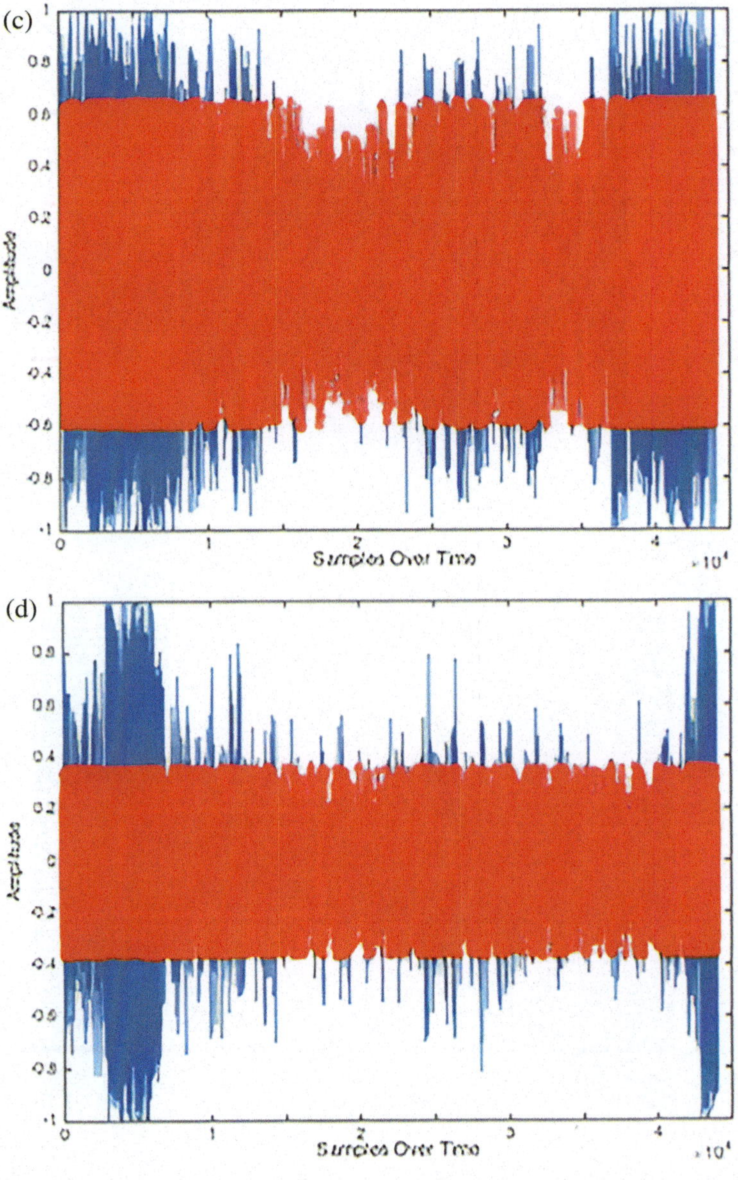

Fig. 2.4 (continued)

different from the loaded ball mill segment. Generally, for the empty ball mill, the amplitude range varies between − 0.6 to + 0.6 and during the transition from one cycle to another, the value is in the range − 0.2 to + 0.2. The block as marked in Fig. 2.5a and b depicting the state of transition.

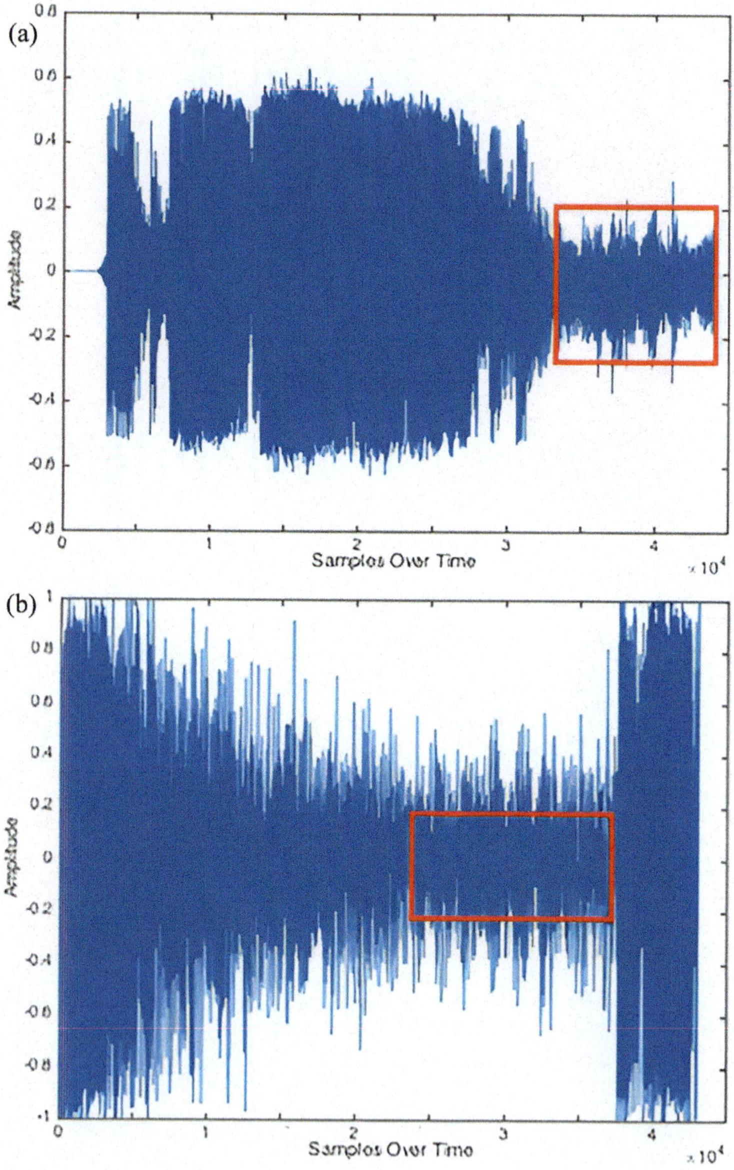

Fig. 2.5 **a** One full cycle of the empty ball mill. **b** Transition of an empty ball mill

(iv) Ball mill datasets are acquired from the time domain and converted into the frequency domain using FFT.

Now, to model a system which can identify different running states of the ball mill, the data is acquired as discussed above and a knowledge base is prepared. The states of the mill are also correlated along with the particle size distribution (psd) of the raw data. For that, after each two minutes of crushing, the materials are brought outside the vessel and the raw material is screened. The experiment is conducted on different materials like Iron, Coal, etc. When the material is Iron, it has been observed that the total crushing time is more or less eight minutes. The total period is divided by two, which generate four stages. These four stages are considered to represent four different crushing situations. For example, the four stages are highlighted as the beginning of crushing, two minutes after crushing, two minutes before the end of crushing ends and the end of crushing.

2.7 State Identification

The full duration of the crushing of a raw material is divided manually into different stages. In an experiment with Iron ore, the total period is considered as 8 min, which is divided into four stages with equal division. There are four stages formed:

(i) Beginning of crushing
(ii) Two minutes after crushing
(iii) Two minutes before the end of the crushing
(iv) End of crushing.

Also, during the operation of an empty ball mill, a state is created named as Empty state.

The operation can be divided into two parts:

(i) Differentiate the running condition of empty and loaded ball mill
(ii) Identification of different stages of running in load-varying conditions.

In this Chapter, the first part is discussed where the empty and the loaded ball mill are identified, whereas the second part will be discussed in Chap. 4.

2.7.1 *Differentiate Empty and Loaded Conditions*

Mel-frequency cepstral Coefficients (MFCC) [17, 18] are widely used for feature representation in audio and speech processing and are known for their ability to capture perceptually relevant characteristics of sound. It is developed to mimic the human auditory system's sensitivity to sound frequencies. MFCC [19, 20] plays a vital role in different applications such as speech recognition, speaker identification, and audio classification.

The core idea behind MFCC is to transform an audio signal into a set of coefficients that summarize its spectral properties. This transformation leverages the Mel scale, a perceptual frequency scale designed to match the way humans perceive pitch. By emphasizing frequencies to which the human ear is most sensitive and compressing

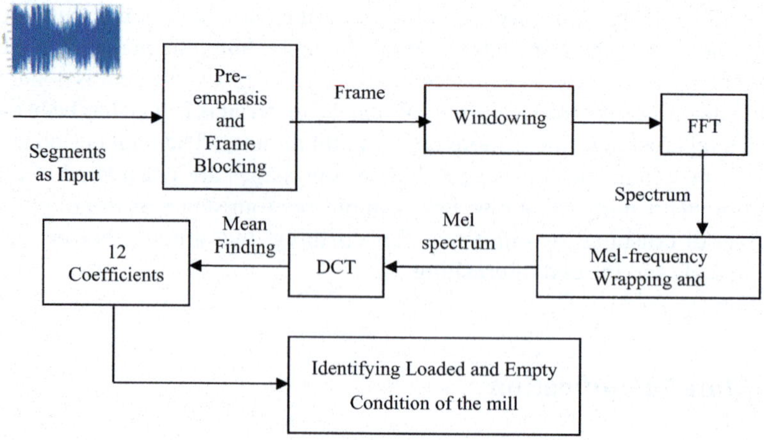

Fig. 2.6 Block diagram of the proposed method for identifying the states

the spectral information into a compact form, MFCC provides a robust representation of audio signals. MFCC has gained popularity because of its ability to reduce the dimensionality of audio data while preserving information critical for distinguishing different sounds. Its mathematical foundation combines signal processing techniques such as the Fourier Transform, Mel-filter bank, logarithmic compression, and Discrete Cosine Transform (DCT), creating a feature set that is both computationally efficient and perceptually meaningful. This makes MFCC [21] an essential building block for many modern audio analysis systems, bridging the gap between raw audio signals and machine learning algorithms.

A mill is running in which condition—empty or loaded can be decided based on those extracted features. However, the different stages of the load-varying conditions cannot be identified like this. That will be discussed in Chap. 4. The block diagram of the method for identifying the empty and loaded mill is shown in Fig. 2.6.

The objective of this method is to distinguish the empty and loaded conditions of the ball mill while running and correlate the corresponding *psd* at each particular state. The ball mill operates in a closed-loop environment, so from outside, it is not possible to predict the *psd* during the running of the mill. Therefore, the aim is to develop a data-driven control system which automatically predicts the *psd* at a particular stage of mill operation by analysing the acoustics of the mill obtained during experiments.

At first, from the knowledge base, the data of transition segments are taken for both the loaded and empty ball mill and are considered as input. The segmented data is stored in a 1-D array for a raw material Iron ore where the crushing operation is completed within eight minutes. The size of samples of each segment is approximately 40–50 K. Each element in the array represents the amplitude at a

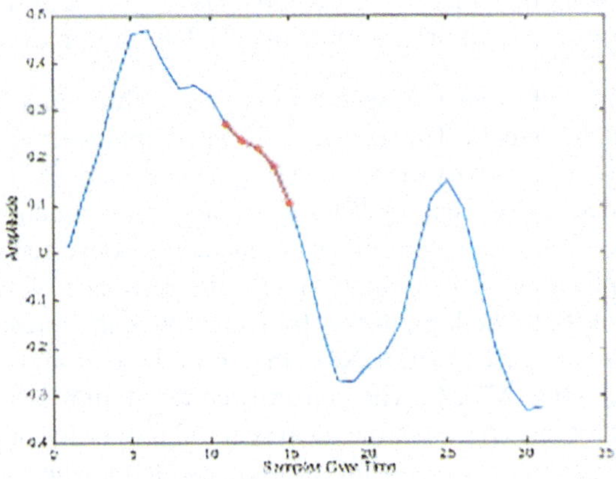

Amplitude values of the red marked area
0.3304
0.2715
0.2362
0.2206
0.1833

Fig. 2.7 Fraction of a segment and amplitude array

particular instant with reference to zero, which may contain a positive or negative value. Figure 2.7 shows a fraction of a segment and the corresponding amplitude array. Following are the different steps followed for MFCC:

(i) Pre-emphasis—In the first phase of the MFCC, the input vector is pre-emphasized to amplify the important high-frequency signal, which was suppressed due to the sound production mechanism of the human auditory system. The audio signal is passed through a high-pass filter and then performs the z-transform on the filtered signal, as given in Eq. (2.1).

$$H(z) = 1 - az^{-1}, \text{ where } 0.95 < a < 0.99 \quad (2.1)$$

(ii) Framing and Windowing—Next, the pre-emphasised output is given as the input for frame-blocking. The amplified signal for ease of processing has been divided into a block of frames consisting of N samples where adjacent frames are separated by M samples such that M < N. Here, M is considered as 256 and N as 100. After frame blocking, hamming windowing is performed to minimise the discontinuities between the first and the last point of a frame. The signal is denoted as $s(n), n = 0, 1, \ldots, N - 1$, and after applying the Hamming window, this signal is represented as $s(n) * w(n)$, where $w(n)$ is the Hamming window defined by Eq. (2.2).

$$w(n, \alpha) = (1 - \alpha) - \alpha \cos \cos\left(\frac{2\pi n}{N - 1}\right), 0 \le n \le N - 1 \quad (2.2)$$

(iii) Fast Fourier Transform—Then the time domain data of each frame is converted into the frequency domain using FFT [22] to obtain the magnitude frequency correlation of each frame. In place of FFT, DFT also can be considered, but due to its simpler approach and time complexity, FFT is chosen over DFT.

Mel Filter Bank—Then, the frequency spectrum is passed through a filter bank designed to emphasize the Mel scale. The Mel scale is a perceptual scale of pitches modelled to reflect human ear sensitivity to different frequency bands. DCT is applied on 20 log energy received after Mel scaling. The frequency domain data is used as input to the Mel-frequency warping where linear frequency is converted into logarithmic frequency. This phase is used to approximate the response of the human auditory systems more closely to the logarithmic frequency than the linearly spaced frequency bands used in the normal cepstrum. Next, the log mel spectrum is converted back to the time domain, called MFCC. The cepstral representation of the speech spectrum provides a good representation of the local spectral properties of the signal for the given frame analysis. The mel spectrum coefficients (and so their logarithm) are continuous, which we convert to the time domain data using Discrete Cosine Transform (DCT). Finally, by applying MFCC on each segment, a matrix of dimension of 256×12 has been generated. The number of columns represents 12 coefficients derived using DCT and used as features. Acoustic parameters or coefficients are represented as in Table 2.1.

Since the segment is divided into 256 frames, for each single coefficient there are 256 values. Each coefficient is represented by the mean value and stored as the MFCC coefficients for each segment. This method reduces the dimensionality of audio data and can be easily integrated with machine learning models. However, this method is not robust to noise; pre-processing steps are often required in noisy environments. It captures only stationary features, which may miss dynamic audio characteristics.

After applying Mel Frequency Wrapping on different segments of the sound of an empty and loaded ball mill, the following results are acquired. The unique pattern has been generated using twelve coefficients for loaded and empty ball mills, shown in Fig. 2.8. Using Fig. 2.8, we distinctly can identify the empty and loaded condition of the ball mill. The states of the loaded mill are divided into three different states, (i) when the crushing starts, (ii) at the middle of crushing and (iii) at the end of crushing, shown in Fig. 2.8. However, excluding these three states, other different load varying conditions are not distinguishable from this approach.

2.7.2 Identification of Different Load-Varying Conditions

Acoustics of the mill under variable load conditions are analysed to identify different stages of grinding manually. At runtime, the proposed method captures the segments of the ball mill *acoustics* under different load-varying conditions. In Fig. 2.8, an output is shown where a ball mill in the laboratory is used for experiments. The Iron ore is taken for the experiment, a certain amount of the input mixture takes 8 min to reach its desired particle size. To perform the experiment, a running ball mill sound

Table 2.1 MFCC parameters

Name of the parameter	Meaning
S	Speech signal
Fs	Sample rate in Hz (default 11,025)
NC	Number of cepstral coefficients excluding 0th coefficient (default12)
N	Length of frame in samples (default power of 2 < (0.03*Fs))
P	Number of filters in filter bank (default: floor(3*log(Fs)) = approx 2.1 per octave)
INC	Frame increment (default n/2)
FL	Low end of the lowest filter as a fraction of Fs (default = 0)
FH	High end of highest filter as a fraction of Fs (default = 0.5)
W	Any sensible combination of the following: 'R' rectangular window in time domain 'N' Hamming window in time domain 'M' Hamming window in time domain (default) 'T' triangular shaped filters in mel domain (default) 'N' Hamming shaped filters in mel domain 'M' Hamming shaped filters in mel domain 'P' filters act in the power domain 'a' filters act in the absolute magnitude domain (default) '0' include 0'th order cepstral coefficient 'E' include log energy 'd' include delta coefficients (dc/dt) 'D' include delta-delta coefficients (d^2c/dt^2) 'z' highest and lowest filters taper down to zero (default) 'y' lowest filter remains at 1 down to 0 frequency and highest filter remains at 1 up to Nyquist frequency. If 'ty' or 'ny' is specified, the total power in the FFT is preserved

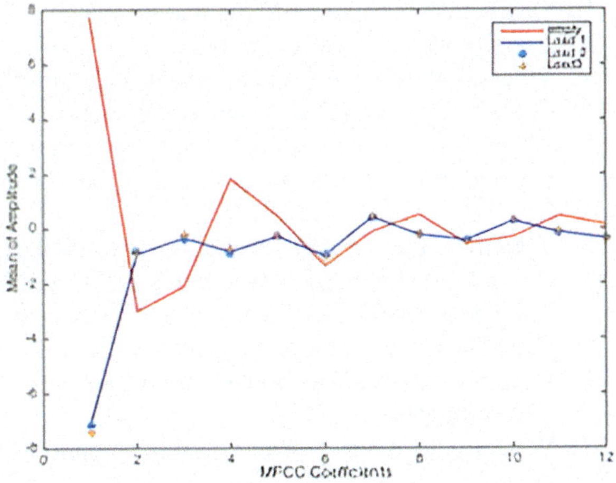

Fig. 2.8 MFCC coefficients of empty and loaded mill

needs to be captured in its different load-varying and empty conditions. Normally, the human ear can easily recognise the changes. To find it automatically, the MFCC is applied on the captured data. After applying Mel Frequency Wrapping on different full-cycle segments including the empty mill and load conditions, the following result is acquired. A unique pattern has been generated for the load and empty condition of the working mill shown in Fig. 2.8. Twelve coefficients are extracted after MFCC from different segments are plotted in Fig. 2.8. From the given Figure, one can easily identify the empty and loaded condition of the ball mill based on the first two coefficients. Later on, the coefficients are overlapped. From this representation, the loaded and unloaded conditions are distinctively identified. However, the states of different load variations are not distinguishable from this approach.

2.8 Summary

The structure of the ball mill acoustics are described in this Chapter. A brief discussion on the time and frequency domain data of the acoustic is explained, along with the procedure of recording and storing sound in the system. A framework has been proposed to identify the working condition of a ball mill by analysing the data. The method distinguishes the loaded mill from an empty mill. Using this concept, the mill is controlled in its closed-loop environment at run time. The model is valid for other different kinds of reactors/crushers/compound machines. In this Chapter, it has been highlighted to distinguish only the empty and loaded ball mills. However, this method is not enough to distinguish the states of the load varying conditions of the ball mill. The next Chapter proposes techniques to identify the different states of the load varying conditions of the ball mill.

Exercise

1. Distinguish between crushing and grinding?
2. Classify various grinding mills based on their working.
3. What are the desirable conditions in a classifier?
4. What is the principle of Magnetic separation?
5. A 100 Hz sinusoid is sampled at S Hz, and the sampled signal is then reconstructed at 420 Hz. What is the frequency of the reconstructed signal if S = 210 Hz?
6. Obtain the Fourier transform of the two sided exponential function $f(t) = e^{\alpha t}$, $t < 0$ and $f(t) = e^{-\alpha t}$, $t > 0$, where α is a positive constant.
7. Find the inverse Fourier transform of $F(\omega) = 20 * (\sin 5\omega / 5\omega)$.
8. Can MFCC be used for non-speech audio analysis? If so, in what applications?
9. How does noise affect MFCC-based speech recognition systems, and how can it be mitigated?
10. What are the main steps involved in computing MFCC features from a speech signal?

References

1. Ajaal, T., Smith, R.W., Yen, W.T.: The development and characterization of a ball mill for mechanical alloying. Can. Metall. Q. **41**(1), 7–14 (2002)
2. Wang, X., Lu, X., Jia, Z., Wang, W., Wang, F.: Control system for micro milling machine. Int. J. Ind. Syst. Eng. (2015). https://doi.org/10.1504/IJISE.2015.070876
3. Powell, M.S., Smit, I., Radziszewski, P., Cleary, P., Rattray, B., Eriksson, K., Schaeffer, L.: The selection and design of mill liners. In: Advances in Comminution, pp. 331–376. Society for Mining, Metallurgy, and Exploration, Inc., Colorado, USA (2006). ISBN-13:978-0-87335-246-8
4. Fedoryshyn, R., Zagraj, V., Pistun, Y.: Automation and optimization of coal grinding by means of ball mills. In: Katalinic, B. (ed.) Proceedings of the 22nd International DAAAM Symposium, vol. 22, no. 1, pp. 0421–0422. DAAAM International, Vienna, Austria, EU (2011). ISSN 1726-9679 ISBN 978-3-901509-83-4
5. Gupta, A., Yan, D.: Mineral processing design and operations. In: An Introduction. Elsevier, p. 718 (2006)
6. Fuerstenau, D.W., Kapur, P.C., De, A.: Modelling breakage kinetics in various dry comminution systems. Kona Powder Part. J. **21**, 121–132 (2003)
7. Spencer, S.J., Campbell, J.J., Weller, K.R., Liu, Y.: Acoustic emissions monitoring of sag mill performance. IEEE Explore (1999). https://doi.org/10.1109/IPMM.1999.791509
8. Campbell, J., Spencer, S., Sutherland, D., Rowlands, T., Weller, K., Cleary, P., Hinde, A.: SAG mill monitoring using surface vibrations. In: Proceedings International Autogenous and Semiautogenous Grinding Technology, pp. 373–385 (2001)
9. Hartmann, W., Candy, J.: Acoustic Signal Processing (2014). https://doi.org/10.1007/978-1-4939-0755-7_14
10. Gong, C.S.A., Lee, H.C., Chuang, Y.C., Li, T.H., Su, C.H.S., Huang, L.H., Hsu, C.W., Hwang, Y.S., Lee, J.D., Chang, C.H.: Design and implementation of acoustic sensing system for online early fault detection in industrial fans. J. Sens. **1**, 15 (2018)
11. Adaileh, W.M.: Engine fault diagnosis using acoustic signals. In: Applied Mechanics and Materials, vol. 295–298, pp. 2013–2020. Trans Tech Publications, Ltd. (2013). https://doi.org/10.4028/www.scientific.net/amm.295-298.2013
12. Zoz, H., Reichardt, R., Kim, J.S.: Application and design of drum mills. In: Proceedings of the World Congress on Powder Metallurgy & Particulate Materials, pp. 16–21. Orlando, FL, USA (2002)
13. Erdem, A.S., Ergun, S.L.: The effect of ball size on breakage rate parameter in a pilot scale ball mill. Miner. Eng. **22**, 660–664 (2009)
14. Deniz, V.: A study on the specific rate of breakage of cement materials in a laboratory ball mill. Cem. Concr. Res. **33**, 439–445 (2003)
15. Fuerstenau, D.W., Lutch, J.J., De, A.: The effect of ball size on the energy efficiency of hybrid high-pressure roll mill/ball mill grinding. Powder Technol. **105**, 199–204 (1999)
16. Sikorski, W.: Acoustic Emission—Research and Applications. InTech (2013). ISBN: 978-953-51-1015-6
17. Kamarulafizam, I., Salleh, S.H., Najeb, J.M., Ariff, A.K., Chowdhury, A.: Heart sound analysis using MFCC and time-frequency distribution. In: 3rd Kuala Lumpur International Conference on Biomedical Engineering, IFMBE Proceedings, vol. 15, pp. 402–405. Springer-Verlag, Berlin, Heidelberg (2007). https://doi.org/10.1007/978-3-540-68017-8_102. ISBN: 978-3-540-68016-1
18. Thu, L.N., Oo, H. N.: Acoustics-based vehicle classification system using MFCC and wavelet-based MFCC. In: 8th International Conference on Science and Engineering (2018)
19. Frigieri, E.P., Brito, T.G., Ynoguti, C.A., et al.: Pattern recognition in audible sound energy emissions of AISI 52100 hardened steel turning: a MFCC-based approach. Int. J. Adv. Manuf. (2017). https://doi.org/10.1007/s00170-016-8748-421

20. Eskidere, Ö., Gürhanlı, A.: Voice disorder classification based on multitaper mel frequency cepstral coefficients features. Comput. Math. Methods Med. **2015**(956249), 12 (2015). https://doi.org/10.1155/2015/956249
21. Singh, S., Rajan, E.G.: MFCC VQ based speaker recognition and its accuracy affecting factors. Int. J. Comput. Appl. **21**(6), 1–6 (2011). https://doi.org/10.5120/2519-3423
22. Harčarik, T., Bocko, J., Masláková, K.: Frequency analysis of acoustic signal using the fast Fourier transformation in MATLAB. Procedia Eng. **48**, 199–204 (2012). https://doi.org/10.1016/j.proeng.2012.09.505

Chapter 3
Particle Size Analysis and Predictive Grinding

Abstract This Chapter aims to describe the stages of developing an appropriate control schematic for dynamic breakage with respect to particle size distribution. The first phase of the work is to establish a proper theoretical background in order to establish the relationship of grinding of one type of particle to another type within a definite size range. Sufficient experimental results obtained from laboratory data, conducted in the Mineral dressing laboratory using a ball mill at Indian Institute of Engineering Science and Technology, Shibpur, India, which supports the theoretical framework. Experimental analysis of the data reveals some interesting results that help to predict the particle size distribution of materials, obtained through grinding operation. A variant of back propagation learning algorithm is used to train the input-output values of different kinds of particles crushed by the ball mill. Instead of a single value, the multi-valued information of the crushed materials, represented using the p-dimensional vector ($p > 1$), is fed to each node of the input-output layer of a feedforward neural network. The training algorithm has been studied for a wide range of input-output values and satisfactory results are obtained, especially when the output vector is small enough compared to the input vector. Experimental data are verified with the simulation results to establish the fundamental relationships between grinding of different kinds of particles.

3.1 Introduction

Grinding circuits are relatively simple and structured. Systems of mechanical process engineering are employed to produce small-size materials at the expense of energy consumption. A lot of work [1–3] has already been reported towards optimization of grinding and distribution of particle size. However, till now difficulties exist in the application of different algorithms to practical problems. The bottleneck involves mainly due to a lack of methods for measuring important state variables, and ambiguity in the determination of model coefficients used for prediction and porting to different particle sizes. One of the intentions of the current contribution is to prove

J. Sil et al., *Intelligent Real-Time Control Systems to Predict the Particle Size of Minerals Using Acoustics of Ball Mill*, Studies in Systems, Decision and Control 600,
https://doi.org/10.1007/978-3-031-91290-0_3

that sufficient experimental results with certain indexes can establish the relationship of grinding of one kind of particle to another kind within a definite size range. This simplification has further significance since some fundamental relationships for several grinding circuit configurations could be obtained from the interpretation of the derived results.

The purpose of grinding is mainly to reduce the particle size and liberation of pure material. However, it should be kept in mind that grinding is not a beneficial operation.

The size to which particles are reduced is governed by three criteria:

(i) The required product size, when simple sizing is desired.
(ii) The particle size of the mineral grains in an ore, which are to be liberated for later separation and collection into a concentrate state.
(iii) Exposure of mineral grains for dissolution of the leaching process.

The determining factor in any breaking operation (fragmentation of particles by blasting to fine grinding) is the particle weight size and its distribution. Grinding parameters of the ball mill are needed to design an expert controller for developing an autonomous grinding system. The conventional way of functioning the ball mill [3] does not provide the mechanism to adapt the grinding parameters. The feed-forward neural network (NN) model is suitable for this purpose where particle size distribution for the product range (target) is predicted [4–6] using the supervised learning algorithm of Rumelhart's [7] back-propagation (BP) algorithm. Here the algorithm has been modified to generalize the algorithm in order to recognize the unseen input mix vectors.

Instead of applying a single input-output pattern at a time and computing the sum squared error (SSE) for each such pair, in the proposed method [4–6], a number of training pattern vectors (input-output pattern) are applied simultaneously at each node of the input layer of a three-layered feed-forward NN. An input vector representing particle size distribution is mapped at an input node, unlike a single element of the input vector in the BP training algorithm. Similarly, target vectors representing the desired particle size distribution are also mapped at every node of the output layer. Using the steepest descent type of iterative approximation [8], the algorithm tries to minimize the SSE over all the output classes. Using the trained weight set of the NN, for input size distribution of test patterns corresponding product size distribution is predicted at the output layer. Since load sensors are used to determine weights, these can be mapped to grinding parameters of the ball mill, which guarantee to generate the desired particle size distribution within a limited size band irrespective of the feed.

In this Chapter the predictive grinding, and experimental set up for obtaining data and analysis of the result at different stages are described. Also the determination of the grinding relationship ('*K*' factor) of one kind of particle to another kind within a definite size range and justification on selecting a random function for this is given afterward. The NN model for predictive grinding, justification of breakage characteristics, and the generalized supervised training algorithm are described after that.

3.2 Predictive Grinding

Designing of an appropriate control schematic for dynamic breakage with respect to particle size distribution is nothing but establishing an appropriate theoretical background adequately supported by laboratory data (for verification and validation vis a vis conceptualization at the initial stage). The main objective is to develop a framework for a smooth transition between different types of grinding using parameters, which are derivable from the established state-of-the-art grinding processes. Moreover, the complexity of the size distribution problem seeks a mechanism that is flexible enough to be tailored to any breakage pattern.

Different scientists have offered different answers to this problem. The major classes of the solution are:

(a) Commonly used functions, such as Rosin-Rammler, Gaudin-Schuman, Broadbent—CallCott, Gaudin—Meloy Equations are to name a few [1, 3].
(b) More complex functions are Svarovsky, Clotz and Schubard functions involving particle physical properties [1, 3].
(c) Matrix notation for size distribution.

However, none of these approaches fits well enough for the transition of one kind of grinding material to another kind.

3.2.1 *Experimental Setup*

The sieve shaker is a stack of seven to eight screen bowls of which the largest aperture is kept at the topmost position and a solid bowl is kept at the lowermost position to collect the finest materials. The screen bowls are stacked from large to small and put inside the shaker assembly, the details of which may vary from make to make. The sieves are tightly placed on its base and the rotary percussive oscillation is put into the assembly to segregate the material into different size fractions.

In industry, generally multi deck screens are used but in the case of grinding in tandem (in series grinding) the particle (feed) after sieving again passes through a large screen. Only the deck along with the ball mill is not sufficient to frame the experimental set-up for testing and validating the hypothesis. Therefore, the work has been phased out into small processes to emulate the real-life scenario. The ideal close circuit diagram of the experiment is shown in Fig. 3.1.

In the experiment, a sieve shaker splits the input feed with different fixed-size intervals while the output can be obtained after classifying the particles into similar clusters. Before crushing, the material is sieved in a sieve shaker for approximately ten minutes. Particles remaining in each sieve denote the particle class between its upper size and lower size which further provides more information: the *class interval*, *particle mass* of that class and the *ratio of class interval to total input* that implies the *mass fraction* with respect to that particular sieve. The materials thus classified are

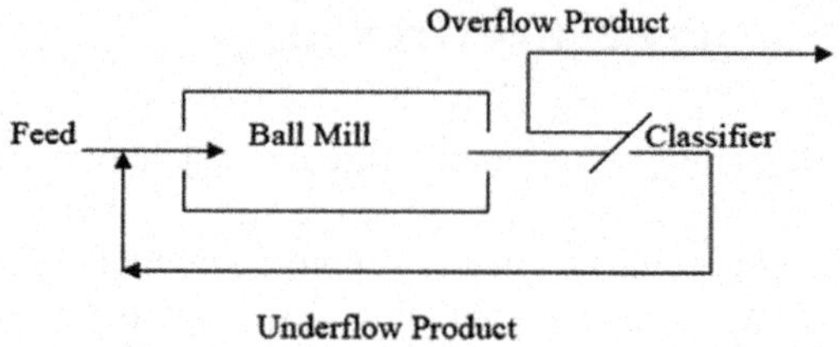

Fig. 3.1 Simple closed grinding circuit

submitted inside the crusher for a given duration, called *residence time.* Since it is a batch process, change of *residence time* results in changes of breakage characteristics which are recorded after each run.

Output obtained from the ball mill is again fed inside the sieve shaker and sieved for a finite duration (10–12 min) to extract the pattern of breakage in each class. Thus, a *mass fraction* of output is also obtained and recorded. The experimental data for classifying input and output patterns are given in Table 3.1.

It is worth mentioning that in industrial applications only the oversized fraction of the first phase (part of feed) is fed into the second stage but here emulation the entire output of the first stage is fed to the second stage to obtain the characteristics of the grinder. Moreover, unlike industry, the undersized material is not consumed by downstream processes. Therefore, the scope of tuning the ball mill at any stage is prevented by the non-availability of oversized and undersized particles obtained

Table 3.1 First stage of the grinding process

Feed size		Product size	
Size	Weight (gm)	Size	Weight (gm)
+ 5.6 (mm)	313	5.6 (mm)	436.9 (with pan)
+ 4.75 (mm)	40	4.75 (mm)	210.8
+ 1.70 (mm)	37	1.7 (mm)	233.4
+ 1.18 (mm)	202.6 (with pan)	1.18 (mm)	202.1
+ 850 μ	203.6	850 μ	19.5 (with paper)
+ 710 μ	197.1	710 μ	11.9
+ 500 μ	203.5	500 μ	22.2
+ 300 μ	206.7	300 μ	27.2
+ 90 μ	218.4	90 μ	49.6
+ 75 μ	200	75 μ	19.2
+ 45 μ	196.5	45 μ	9.8

Weight of ball: 2.7 kg
Revolution time: 5 min
Weight of pan: 195 gm
Weight of paper: 8.7 gm

from the experiment and hence the variation from the anticipated characteristics of breakage can be ascertained only from the anomaly, either at the input or at the intermediate stages. To overcome the constraint, the blend of oversized and undersized and hence the *mass fractions* are kept intact in case of both primary and secondary grinding.

In grinding, primary grinding and secondary grinding refer to different stages of material size reduction in mineral processing, cement manufacturing, and other industries that require fine particle grinding. Primary grinding aims at reducing large feed particles often within the ranges from 25 to 100 mm to an intermediate size, and usually produces coarser particles, around 1 to 5 mm. However, in the secondary grinding the particle further reduces to achieve a finer, more uniform size and produces fine particles, often below 100 microns. In short, primary grinding breaks down material to an intermediate size, while secondary grinding achieves finer particle sizes for better material processing and extraction.

3.2.2 *Experimental Analysis*

The work follows the standard procedure using a sieve shaker and multistage grinding machine (ball mill in tandem) for the experiment. The first part of the work derives a solution out of the laboratory data obtained from the mineral dressing laboratory, Indian Institute of Engineering Science and Technology, Shibpur, India which has been described below.

First Stage of Grinding

The feed size (input) and product size (output) obtained from the first stage of the multistage grinding process are given in Table 3.1.

Second stage of Grinding

The feed size (input) and product size (output) obtained from the second stage of the multistage grinding process are given in Table 3.2.

Multi Stage Grinding

The tasks of the multistage grinding process have been summarized below.

(a) Selection of appropriate amount of feed.
(b) Size classification of feed.
(c) Fed the charge size distribution in the ball mill.
(d) *Residence time* is 10 min.
(e) Size classification of crushed product.
(f) Repeat process (a) to (e) for other materials.

First Stage: Feed size distribution should be the same with the initial feed size.

The corresponding details of Feed Size, Mass and the Mass fraction is given in Table 3.3.

Table 3.2 Second stage of grinding—feed product size distribution

Feed size		Product size	
Size	Weight (gm)	Size	Weight (gm)
+ 5.6 (mm)	436.9 (with pan)	5.6 (mm)	420.3 (with pan)
+ 4.75 (mm)	210.8	4.75 (mm)	211
+ 1.70 (mm)	233.4	1.7 (mm)	218.9
+ 1.18 (mm)	202.1	1.18 (mm)	198.8
+ 850 μ	19.5 (with paper)	850 μ	200.2
+ 710 μ	11.9	710 μ	10.1 (with paper)
+ 500 μ	22.2	500 μ	18
+ 300 μ	27.2	300 μ	41
+ 90 μ	49.6	90 μ	85.8
+ 75 μ	19.2	75 μ	17.6
+ 45 μ	9	45 μ	9

Weight of ball: 2.7 kg
No. of revolution: 182
Revolution time: 4.45 min
Weight of pan: 195 gm
Weight of paper: 8.7 gm
R.P.S.: 50

Table 3.3 First stage of multistage grinding—feed size distribution

Feed size (dp)	Mass (gm)	Mass fraction
+ 5.60 mm	313	0.683108
+ 4.75 mm	40	0.087298
+ 1.70 mm	37	0.080751
+ 1.18 mm	(202.6 − 195) = 7.6	0.016587
+ 850 μ	(203.6 − 195) = 8.6	0.018763
+ 710 μ	(197.1 − 195) = 2.1	0.004583
+ 500 μ	(203.5 − 195) = 8.5	0.018551
+ 300 μ	(206.5 − 195) = 11.5	0.025098
+ 90 μ	(218.4 − 195) = 23.4	0.051069
+ 75 μ	(200 − 195) = 5	0.010912
+ 45 μ	(196.5 − 195) = 1.5	0.003274
− 45 μ	(195 − 195) = 0	0
Total	458.2	

First stage Product size (Cylindrical) distribution is given below in Table 3.4 with information particle size, Mass, Mass Fraction and Cumulative Mass Fraction. Pictorial representation of the corresponding Particle Size Distribution is shown in Fig. 3.2.

Second Stage: Feed size distribution

The corresponding details of Particle Size, Mass, Mass fraction and Cumulative Mass Fraction is given in Table 3.5. Pictorial representation of the corresponding Particle Size Distribution is shown in Fig. 3.3.

Table 3.4 First stage of multistage grinding—product size distribution

Particle size (dp)	Mass (gm)	Mass fraction	Cumulative mass fraction (Φ)
+ 5.6 mm	241.9	0.602	0.602
+ 4.75 mm	15.8	0.039	0.641
+ 1.70 mm	38.4	0.095	0.736
+ 1.18 mm	7.1	0.018	0.754
+ 850 μm	10.8	0.027	0.781
+ 710 μm	3.2	0.008	0.789
+ 500 μm	13.5	0.034	0.823
+ 300 μm	18.5	0.046	0.869
+ 90 μm	40.9	0.1	0.969
+ 75 μm	10.5	0.026	0.995
+ 45 μm	0.9	0.002	0.997
− 45 μm	0	0	0.997
Total	401.5		

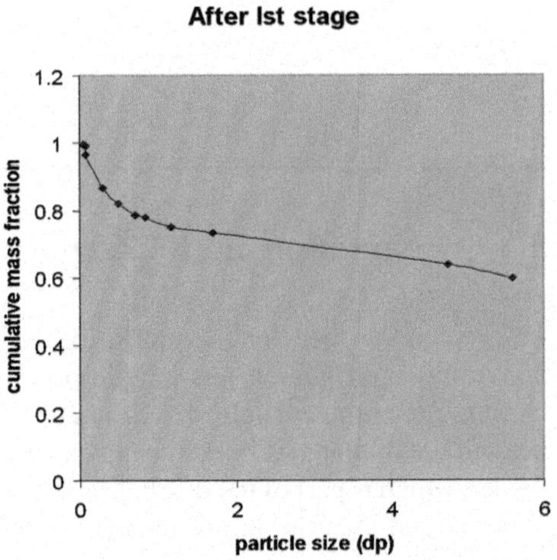

Fig. 3.2 Graphical representation of particle size distribution for iron ore

Table 3.5 Second stage of multistage grinding—feed size distribution

Particle size (dp)	Mass (gm)	Mass fraction	Cumulative mass fraction (Φ)
5.6 mm	220.8	0.55	0.55
+ 4.75 mm	16	0.04	0.59
+ 1.70 mm	23.9	0.06	0.65
+ 1.18 mm	3.8	0.009	0.659
+ 850 μm	5.2	0.013	0.672
+ 710 μm	1.4	0.004	0.676
+ 500 μm	9.3	0.023	0.699
+ 300 μm	32.3	0.080	0.779
+ 90 μm	77.1	0.19	0.969
+ 75 μm	8.7	0.02	0.989
+ 45 μm	0.3	0.001	0.990
− 45 μm	0	0	
Total	401.5		

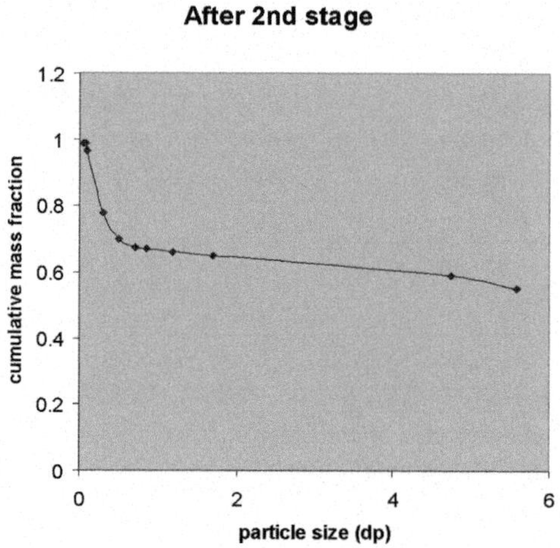

Fig. 3.3 Graphical representation of particle size distribution for iron ore

3.3 Observation on '*K*' Factor

To prepare the feed for the ball mill, the material is sized in a multi deck screen and dozed appropriately. As a rapid prototype it can also be achieved using a sieve shaker. The entire particle size can be sequenced into segments, called bands while the amount of material in each band can be varied by changing the speed of the spiral feeder, which is part of the dozer. Load sensors accurately sense the amount of dozed material and thus the initial feed particle size distribution is available. These data points are sent to the processing element where blended material (feed) is sensed

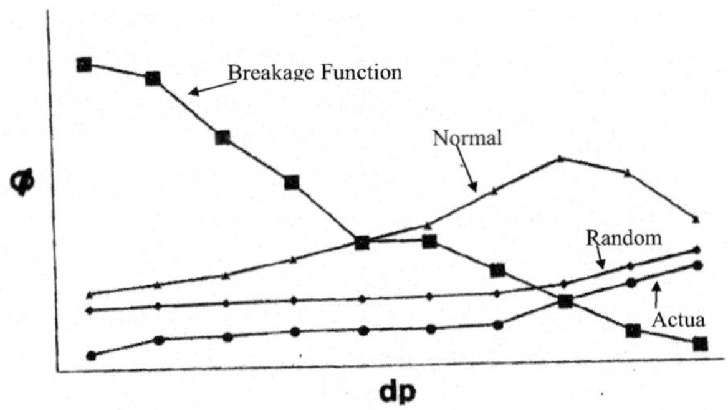

Fig. 3.4 Graphical representation of simulation results using various breakage function

and processed and the resulting data is sent again to the processing element. The difference between the former and the latter is the loss of material on transit and should be rectified according to the proportion of material taken in each band of feed.

It is obvious that multiple changes in the state variables yield different results. In the experiment a linear region has been identified, which is almost parallel or maintains a steady gradient with the horizontal axis, i.e. particle size. Another experimental result shows the relationship with the given particle size range and different hardness of the particle, generates a curve of different vertical span but maintains similarity in horizontal span with the previous one. It has been observed that multiplication by a scalar '*K*' factor enables the human expert to predict the linear region [4–6] for the 2nd set.

The following graphs, given in Fig. 3.4 are obtained by applying the various breakage functions, mentioned with the graphs while the experimental data and the result of simulation select *random function* as the appropriate breakage function.

Thus, it has been concluded that the particle size distribution varies with the mass fraction [4–6] (see, Fig. 3.4) and a multiplication factor (*K*) can de facto give an idea about the change of particle size distribution with the change of different size range within a limited size band. This itself is a contribution to the subject through which one can obtain predictive grinding and particle size distribution using simulation.

3.3.1 Justification of Selecting Random Function

The curves of Fig. 3.4 are drawn from the data obtained after conducting a series of experiments for different input, input particle size distribution, speed of the mill, inclination and grinding media etc. to name a few. The results led to the researcher to conduct a number of simulation with the pre-existing breakage function [1, 3] where different standard mathematical formulae are applied for obtaining a graphical representation that characterize the major portion of breakage with respect to the *mass fraction*. The simulation results are obtained using a number of statistical distribution functions of which random function suits best for the major portion of particle breakage shown in Fig. 3.4. Similar particles of same hardness but different input size distribution can also be correlated with completely different output using a multiplication factor '*K*'. The value of '*K*' de facto implicates the switching characteristics of different patterns of particle breakage.

3.4 Neural Network Model

The neural network (NN) [9] has been used to adapt the dynamic breakage characteristics of the ball mill and to verify the random nature of the breaking process. First of all, to obtain the desired particle size range at the output of the ball mill, the back-propagation learning algorithm has been invoked to train the feed-forward NN for generalization purposes. The parameters of the ball mill are tuned and ready to run for achieving the predicted output. No further training is required so long the desired patterns are unchanged, irrespective of the input particle size range.

3.4.1 Justification of Breakage Characteristics

A learning scheme has been proposed for obtaining feed size distribution and justification [10] of linear zone of particle size distribution that ultimately proves the concept of '*K*' factor, described in Sect. 3.3. A new supervised training algorithm [4–6] using feed forward neural network has been employed where instead of considering a single pair of input-target data in a training set, multi-valued information in the form of a p-dimensional vector ($p > 1$) are submitted at each node of the input-output layer of the network. The particle size distribution within a given size band forms the n-dimensional input vector mapped at each input node while desired particle size distribution within a different specific size band constructs the n-dimensional target vector and mapped at each output node of the NN model. Weights between different layers are adjusted using gradient descent approximation method [8] in order to minimize the SSE at each output node. The training algorithm has been studied for a wide range of input-target pattern vectors and obtaining worthy results, especially when the elements of the target vectors are small enough compared to the same of the input vectors. Since after crushing the particle size becomes smaller than the original ones, the result verifies our assumption.

3.4.2 Learning Scheme

Matrix concept has been introduced in predictive grinding to validate the learning scheme. Let f and m are the feed and the desired (target) product of particle size mapped as the input and output of the NN, respectively. To represent the total breakage process with respect to the mass balance, the following Equations are summarised.

$$f_1 = w_{11} * m_1 + w_{12} * m_2 + w_{13} * m_3 + \ldots\ldots\ldots\ldots + w_{1n} * m_n$$
$$f_2 = 0 + w_{22} * m_2 + w_{23} * m_3 + \ldots\ldots\ldots\ldots + w_{2n} * m_n$$

$$f_q = 0 + 0 + 0 + \ldots\ldots\ldots\ldots + w_{nn*} m_n$$

Hence,

$$F = W * M \tag{3.1}$$

where $F = [f_1 f_2 f_3 \ldots\ldots\ldots f_{\mathrm{q}}]^{\mathrm{T}}, M = [m_1\ m_2\ m_3 \ldots\ m_{\mathrm{n}}]^{\mathrm{T}}$ and w_{ij}, is the component of the matrix, W mapped as the grinding parameters of the ball mill.

W of Eq. (3.1) is an upper triangular matrix, which validates the fact that the breakage process must hold the conservation of mass and also validates the principle that on breakage a larger particle breaks into smaller ones. Moreover, at the state of convergence, the weight matrix remains an upper triangular one, i.e. a band matrix, validating Eq. (3.1).

3.4.3 Supervised Learning Model

A standard three layered feed forward neural network is shown in Fig. 3.5. The network has N input neurons, M middle layer neurons and J output neurons where input signal $X_{\mathrm{p,i}}$ (for $i = 1$ to N) is an input vector of dimension p (where $p > 1$). The weight matrices $W_{\mathrm{N,M}}$ and $U_{\mathrm{M,J}}$ are mapped as input (feed) size distribution parameters [4–6] and output (product) size distribution parameters, respectively.

In the proposed algorithm, since the network is trained with the application of p-exemplar pairs at one shot, the learning is more generalized. Moreover, the algorithm shows good results for training with target values small enough compared to the input. Initially, the weights are randomly set within [0,1] while during training the weights are adjusted to minimise the SSE between the computed output $Z_{p,J}$ and the target output $T_{p,J}$. Following unipolar sigmoidal functions are used as activation functions in the middle and the output layer, respectively.

$$f(r) = 1/(1 + \exp(-a1 * r + b1)),$$

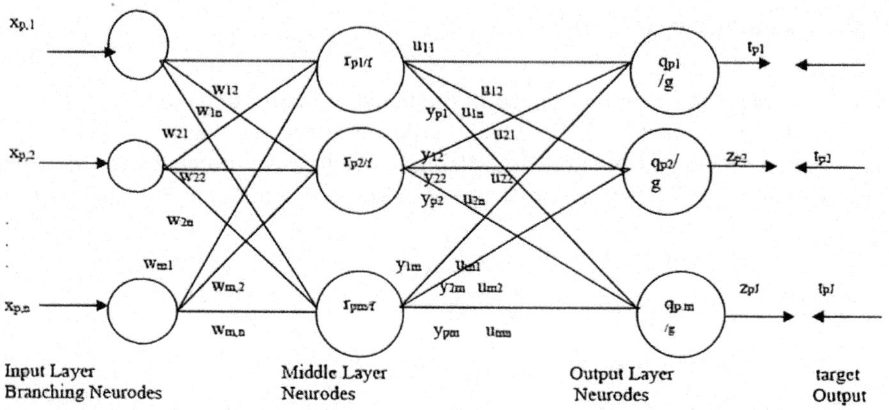

Where i=1,2,3 ... N
l=1,2,3 ... M
j=1,2,3 ... J
and

r=Σwx, where r and z are row vectors of dimension p
y=f(r)
z=g(Σuy)

Fig. 3.5 Neural network model for generalized learning

$$g(q) = 1/(1 + \exp(-a2 * q + b2))$$

where $b1$ and $b2$ are biases, $a1$ and $a2$ are the rise factors. Sum of product of the respective weights and incoming signals at the middle and output neurons are represented by r and q, respectively.

The weights are adjusted using gradient descent method [8]. At each iteration, $U_{M,J}$ and $W_{N,M}$ are updated using Eqs. (3.2) and (3.3), respectively.

$$U_{M,J} = U_{M,J} + \left(T_{p,J} - Z_{p,J}\right) * \left(1 - Z_{p,J}\right) * Z_{p,J} * Y_{p,M} \tag{3.2}$$

$$W_{N,M} = W_{N,M} + \left(T_{p,J} - Z_{p,J}\right) * \left(1 - Z_{p,J}\right) * Z_{p,J} * U_{M,J} * Y_{p,M} * \left(1 - Y_{p,M}\right) * X_{p,N} \tag{3.3}$$

where $Y_{p,M}$ represents the output at the middle layer.

3.4.4 Training Algorithm

Inputs:

N number of input neurodes; *M* number of middle neurodes; *J* number of output neurodes, *N* number of input exemplar vectors $X_{p,N}$, *J* number of target vectors $T_{p,J}$ Termination condition E.

Outputs:

Weight matrices $W_{N,M}$ and $U_{M,J}$ at the hidden layer and output layer respectively.

```
step 1: Read N, M, J, p and exemplar input vectors Xp,N and corresponding target vectors Tp,J
step 2: b1=(N/2)+1; b2=(M/2)+1;
    a1=2.46; a2=2.46;
    E = 0.001 //termination condition
    MNI = 80 // maximum iterations = 80;
    I=0;
step 3: //Initialize wt. matrices W and U by random numbers
    // between 0 & 1.
    for L=1 to M do
      for L1=1 to N do
        WL1,L = random(1);
          for L2=1 to J do
            UL,L2 = random(1);
step 4: //Update neurodes & compute SSE (sum squared error)
     SSE = 0 // initialize sum squared error as 0
   for p1 = 1 to p do
     for M1 = 1 to M do
         R(p1,M1)=0; //where each R(p,i) is a vector of order p
           for p1=1 to p do
             for M1=1 to M do
                for N1 = 1 to N do
                     R(p1,M1) = R(p1,M1) + W(N1,M1) * X(p1,N1);
                      Y(p1,M1) = f(R(p1,M1));
                        //for each output neurodes
                      for p1=1 to p do
                        for J1 = 1 to J do
                          Q(p1,J1) = 0;
                            for p1=1 to p do
                            for J1=1 to J do
                                 for M1=1 to M do
```

```
                    Q(p1,J1) = Q(p1,J1)+U(M1,J1)*Y(p1,M1)
                      Z(p1,J1) = g(Q(p1,J1));
                    for p1 = 1 to p
                      for J1 = 1 to J
                      SSE = SSE + (T(p1,J1) - Z(p1,J1))^2 ;
                      //SSE stores sum of squared errors for all
                      //vectors
step 5: // termination condition
    if (SSE < E) or (I > MNI) then stop  // E is constant having small
    else I = I+1 // small value, MNI is maximum iterations to be performed
step 6: // adjust all weights via formulas
    sum = 0
      for M1=1 to M do   // adjustment of weights of U,M,J
        for J1=1 to J do
          for p1=1 to p do
            sum = sum + (T(p1,J1) - Z(p1,J1)) * Z(p1,J1) * (1-Z(p1,J1)) * Y(p1,M1)
              U(M1,J1) = sum + U(M1,J1);
                for N1 = 1 to N   // adjustment of weights of WN,M
                  for M1 = 1 to M
                  for p1 = 1 to p
                    for J1 = 1 to J
              W(N1,M1) = W(N1,M1)+(T(p1,J1)-Z(p1,J1))*Z(p1,J1)*(1-Z(p1,J1))*
                    Y(p1,M1) *(1-Y(p1,M1))*X(p1,N1)*U(M1,J1);
        goto step 4.
```

The convergence of the algorithm has been successfully tested for a wide range of data collected from the ball mill in operation and especially gives worthy results when target value is much lower compared to input data. However, it is certain that further training would be needed in case the desired particle size resides outside the range.

3.4.5 *Analysis of the Training Algorithm*

The algorithm has been tested for actual experimental data and as well as with simulated data and for both the cases satisfactory results have been achieved, which is described below.

(a) Analysis with Experimental Data

Table 3.6 Summarized output of mass fraction of product

Feed (mass fraction)	Product (mass fraction)
0.683108	0.55
0.087298	0.04
0.080751	0.06
0.016587	0.009
0.018763	0.013
0.004583	0.004
0.018551	0.023
0.025098	0.080
0.051069	0.190
0.010912	0.020
0.003274	0.001

The data fed to the simulation program is real life grinding data obtained from an actual experiment of the ball mill. The input has been prepared from the feed data, obtained from the column of mass fraction of Table 3.3 and the target output of the algorithm has been prepared from the column of mass fraction of the product data of Table 3.5, as summarized in Table 3.6.

It has been observed that the computed output is very satisfactory and has converged to zero percent error, compared with target (desired) output within the residence time (5–10 min). The convergence of error with time is shown in Fig. 3.6.

(b) Analysis with Simulated Data

Input and target output (desired) have been generated using random function, given in Table 3.7. Though the data are random, but are chosen in keeping with similarity to the actual experimental data, maintaining the property of mass fraction, i.e. the overall mass fraction should not go beyond unity.

The convergence of error with time using simulated data (Table 3.7) applied on the NN model is shown in Fig. 3.7.

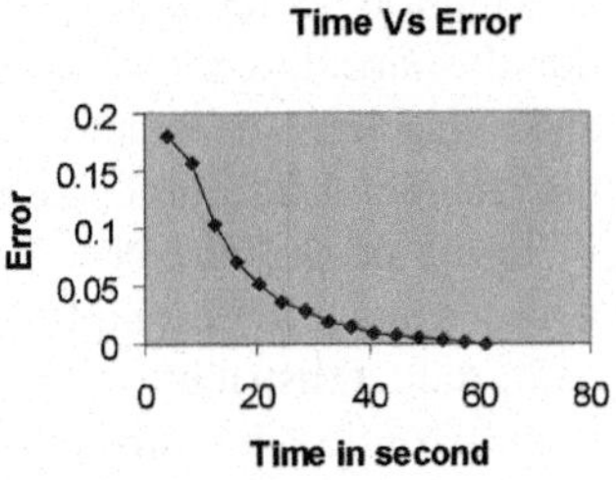

Fig. 3.6 Error convergence using experimental data of Table 3.6

Table 3.7 Initialization of input and target output (desired) using random function

Feed data (input to first stage): mass fraction					Target (desired output) data: mass fraction				
0.092	0.000	0.000	0.000	0.000	0.060	0.000	0.000	0.000	0.000
0.096	0.075	0.000	0.000	0.000	0.069	0.076	0.000	0.000	0.000
0.096	0.068	0.047	0.000	0.000	0.050	0.066	0.072	0.000	0.000
0.099	0.071	0.046	0.024	0.000	0.046	0.065	0.076	0.071	0.000
0.096	0.065	0.046	0.028	0.013	0.062	0.079	0.086	0.084	0.070

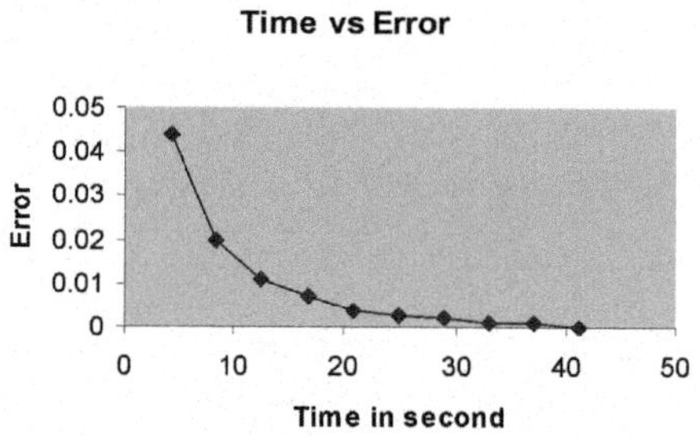

Fig. 3.7 Error convergence using simulated data of Table 3.7

3.5 Summary

In a ball mill, the exact state of breakage of the particle is quite unpredictable since each particle breaks in a different manner. This Chapter describes how the information about particle size distribution during grinding is utilized effectively to establish a relationship between different kinds of grinding materials that helps predictive grinding. Analyzing the breakage characteristics of different materials (coal, quartz etc.) along with their physical properties like size, diameter and specific gravity, a relationship has been established that determines a multiplication factor '*K*' of one kind of grinding to another kind within a linear size range. The prediction of particle size distribution takes place in the linear region, which has been verified through experimental results. The experimental result has been tallied with the simulation result obtained using the proposed NN model where it has been observed that random distribution scores fairly well within a limited size band. It is therefore concluded that the particle size distribution varies with the mass fraction (see, Figs. 3.2 and 3.3) and a multiplication factor '*K*' de facto gives an idea about the change of fines with the change of different size range within a limited size band.

In the proposed NN model, the trained weight sets between input-middle layer neurons and middle-output layer neurons (see, Fig. 3.5) represent the amount of feed and product distribution, respectively. The non-linearity of breakage characteristics of the particle size distribution has been verified using the NN model. The generalized supervised learning scheme also helps to pruning of unimportant weights/grinding parameters, which less likely affect the operation of the ball mill.

Exercise

1. What is Multistage Grinding?
2. Why do we need Random Function for breakage function?
3. What is the 'k' factor?
4. In the back propagation algorithm for the feed forward Neural Network Model, what method is generally used for achieving global minima? And How do we achieve it?
5. Is there any material loss if the residence time is increased during particle grinding? If yes then will that be an undersized or oversized loss? Justify.

References

1. Gills, W.A.: Mineral Processing. Springer Verlag, Germany (1991)
2. Taggert, H.: Hand Book of Mineral Processing. SME Publishers (1951)
3. Wills, B.A.: Mineral Processing Technology. Pergamon Press, UK (1986)
4. Bhaumik, A., Sil, J., Banerjee, S.: Hybrid control of tumbling mill using NN based schematics. Int. J. Inf. Comput. (IJICS) **2**(1), 21–28 (1999)
5. Bhaumik, A., Sil, J., Banerjee, S.: A new supervised training algorithm for generalized learning. In: Third International Conference on Computational Intelligence and Multimedia Applications, ICCIMA '99, pp. 489–493 (1999)
6. Bhaumik, A., Sil, J., Banerjee, S., Dutta, S.K.: Supervised learning algorithm for open loop NN based control of tumbling mill. In: TENCON 99, Proceedings of the IEEE Region 10 Conference, vol. 1, pp. 395–398 (1999)
7. Rumelhart, D.E., Hinton, G.E., Williams, R.J.: Learning Internal Representations by Error Propagation in Parallel Distributed Processing, vol. 1, pp. 318–362. MIT Press, Cambridge, MA (1986)
8. Looney, C.G.: Stabilization and speedup of convergence in training feedforward neural networks. Neurocomputing **10**, 7–31 (1996)
9. Philip, W.D.: Neural Computing Theory and Practice. Van Nostrand Publisher (1989)
10. Zhou, W.: Verification of the nonparametric characteristics of backpropagation neural networks for image classification. IEEE Trans. Geoscience and Remote Sensing **37**(2), 771–779 (1999)

Chapter 4
State Prediction of a Running Ball Mill

Abstract This Chapter aims to propose two efficient strategies to predict the states of a running ball mill in different load varying conditions. The first method is to establish a mathematical model to obtain the acoustic signatures or patterns of different states of transitions of the running ball mill by analyzing the sound of the ball mill in its load varying conditions. In this approach, using Kernel Density Estimator (KDE) and curve fitting algorithm simulated patterns of the acoustics are obtained. On the other hand, in experimentation, the ball mill is stopped manually at the fixed time interval representing the states and the particles are collected and corresponding particle size distribution (*psd*) is calculated. In order to develop the mathematical model which predicts the acoustic signature corresponding to the states, errors between experimental and simulated patterns at different instant of time are calculated and minimized by learning the parameters of the model. The second method predicts the terminating condition of the ball mill by extracting statistical features using time domain information of the acoustics like amplitude, frequency and phase. Using principal component analysis (PCA) method, important features are selected and the data points are clustered into two groups using K-means clustering algorithm representing the End state and Start state of the ball mill. Finally, a Naive Bayesian classifier is trained to predict the termination condition of the running ball mill and when stopped the desired *psd* of the crushed material as targeted by the user has been obtained.

4.1 Introduction

In Chap. 2, the acoustic signal of the running ball mill was analysed to identify the "Empty" and "Loaded" conditions of the mill. The a*coustics* of the mill under variable load conditions are analysed in this Chapter and correlated with the particle size distributions (*psd*) of the crushed particles. This Chapter also aims to predict the terminating criteria to stop the mill when the desired *psd* of the crushed material as targeted by the user has been achieved. At runtime, the proposed method captures

J. Sil et al., *Intelligent Real-Time Control Systems to Predict the Particle Size of Minerals Using Acoustics of Ball Mill*, Studies in Systems, Decision and Control 600,
https://doi.org/10.1007/978-3-031-91290-0_4

the ball mill *acoustics* under different load-varying conditions and transforms to corresponding data distribution.

The experiments are conducted in the laboratory, and after a fixed time interval, say, one minute, the mill stops to collect the grinded materials and obtain *psd* as the outcome of the crushing procedure. As discussed in Chap. 2, during grinding, different parameters like—raw material weight, mill speed, grinding media type and weight, expected *psd* etc. are decided based on the type of the raw material and the expected *psd*. A knowledge base is prepared to store the information about the *psd* of the raw material and other necessary parameters of grinding during each minute of crushing in order to distinguish the different states of the mill operation. The ball mill acoustics are captured continuously as the crushing is continued and stored as *.wav [] format. Continuous acoustic is then divided into segments with respect to four different running states including the terminating state.

The segmented signals captured at different time intervals are transformed into the time domain signal to obtain the raw data. Unique patterns of each different state of crushing are derived using Kernel Density Estimator (KDE) function [1]. Next, Gaussian curve fitting algorithm [2, 3] has been applied to fine tune the patterns. Errors between the experimental and simulated patterns at different states are minimized by developing a mathematical model.

Except the common features of an acoustics like amplitude, frequency and phase there are few more features of the ball mill acoustics which are extracted by calculating statistical moments from the segmented raw dataset. The PCA [4, 5] method is applied to select the most important features as principal components. Based on the selected features, the Start and End state of the ball mill has been labeled by clustering the data points using K-means algorithm [6]. The Naive Bayes classifier [7] is trained using the uniform samples of the clusters and finally, based on a new input feed, the mill has been stopped automatically and the desired *psd* of the materials are collected and verified with the knowledge base.

4.2 Acoustic Signature Based Modelling

The first part of this Chapter contributes towards predicting different states of the running ball mill under its different load-varying conditions which correlates the *psd* of the material. The proposed method generates unique patterns of the mill acoustics using KDE. Few works are reported [8, 9], where the KDE method is used for control system generation based on acoustic emission. KDE is also applied to analyse the ECG signal [10]. In another work [11], the authors discussed how to apply Kernel Density on large data to maintain the quality and efficiency of the data.

In the proposed framework, while experimenting, the continuous acoustics of the mill are captured and the ball mill is stopped manually at fixed time intervals. After a certain instant of time of running, the mill has been stopped and the material of the mill vessel is screened using a sieve shaker to find the psd after crushing and stored in the knowledge base prepared for this experiment. The different grinding parameters like—mill speed, grinding media information, input mix *psd* etc. as discussed in Chap. 2, also need to be stored within the knowledge base. This procedure of preparing the knowledge base and the running ball mill sound capture is continued till the expected fines are generated. Next, KDE function is applied on the dataset where parameters are initialized with the help of the paper [12]. The parameters are tuned using the Gaussian curve fitting method. Patterns are represented as unique signatures of the acoustics of the running ball mill. Required error minimization is applied to follow the experimental result. The workflow diagram of the proposed method is given in Fig. 4.1.

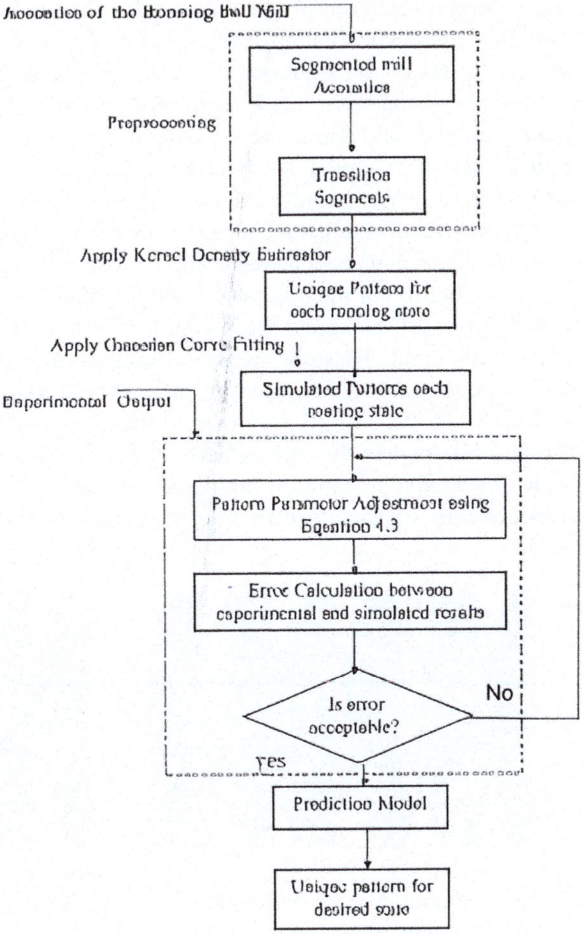

Fig. 4.1 Flow diagram of acoustic signature based modelling

4.2.1 Preprocessing of Method 1

The aim of the Chapter is to identify different states of the running ball mill during the grinding operation of any raw material. First, the acoustics of the running ball mill under different load varying conditions are captured continuously with an input mix till the expected fines are produced and stored in the **.wav* format. Along with the sound information of the running ball mill, the particle size distribution (*psd*) of the input mix are obtained after each specific interval of time where the input mix is screened using a sieve shaker to find the new psd of the material after a certain time of crushing. Now the captured sound is segmented based on one full vessel rotation and the transition from one cycle to another. Figure 4.2 shows the segment representation for both full mill cycle and the transition cycle within the continuous acoustics using red box (discussed elaborately in Chap. 2). Then the segmented cycles are transformed in the time domain data and stored as raw data in vector representation. Each entry of the vector represents the amplitude, plotted against time. Sound waves are typically represented as oscillations around a central equilibrium point (zero), so the amplitude value of the sound can be negative or positive in reference to that equilibrium point. An example of a time series array and its corresponding fragmented signal is shown in Fig. 4.3.

There are two types of acoustic segments—full cycle which is the segment to one complete rotation of the mill vessel and the other cycle is the transition cycle which is the portion of the sound that is the transit time between two consecutive cycles of the mill vessel. These two sets of segments are stored in two different databases to prepare the knowledge base for the proposed model. The first database with full mill cycles is named Full_Cycles, and the other one is named Transition_cycles. It is noticed that when the ball mill completes one full cycle and moves on to the next cycle, the change of sound is highly recognisable. Here, the transition segments are considered for data analysis in order to predict the states. The mill that is used in the

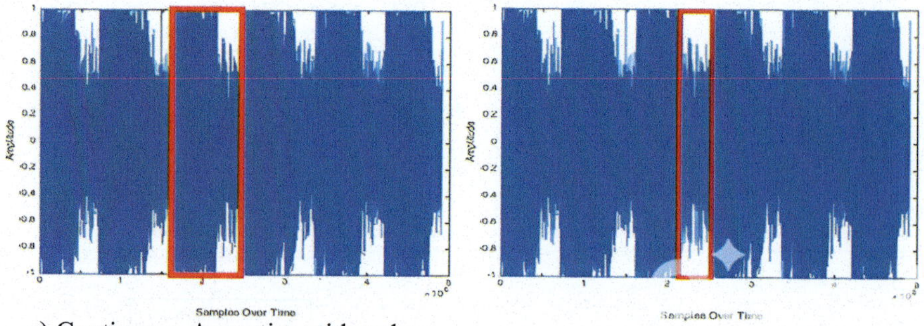

a) Continuous Acosutics with red coloured segment as full cycle

b) Continuous Acosutics with red coloured segment as transition cycle

Fig. 4.2 Segments from the continuous acoustics

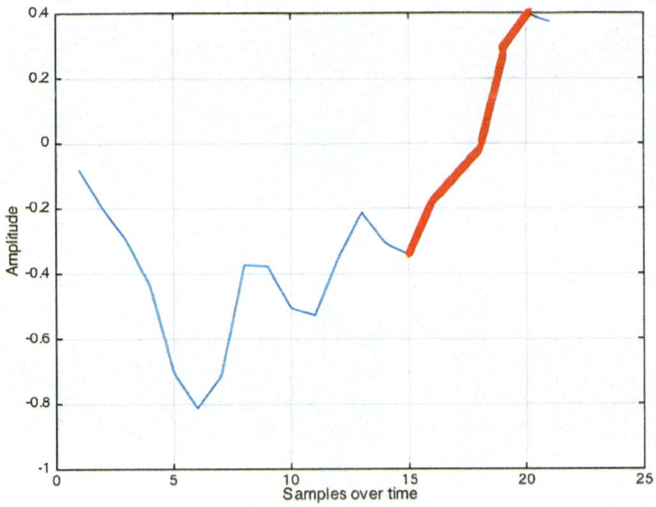

Amplitude of the red marked portion
-0.19207
-0.11724
-0.01535
0.264770
0.399749
0.374481

Fig. 4.3 Amplitude array for the highlighted portion

laboratory for the experiment, operated for four seconds to complete one full cycle and two seconds to transit from one full cycle to the next. Table 4.1 shows the partial information of one of the databases, i.e. Full_Cycle and Table 4.1 shows the other one, i.e. Transition_Cycle Database.

The transition segments are now divided into four different states from the beginning till the end of crushing. Say, within T minutes, the input mix completes its grinding and produces the required fine. Then, the total period is divided into four

Table 4.1 Database information of acoustic segments/cycles

Time instant	Cycle	Range of amplitude
Full_Cycle		
1	Full_Cycle_1	[− 0.8 to + 0.8]
2	Full_Cycle_2	[− 0.789 to + 0.78]
:	:	:
N	Full_Cycle_N	[− 0.35 to + 0.35]
Transition_Cycle		
1	Trans_Cycle_1	[− 0.8 to + 0.8]
2	Trans_Cycle_2	[− 0.6 to + 0.6]
:	:	:
N	Trans_Cycle_N	[− 0.2 to + 0.2]

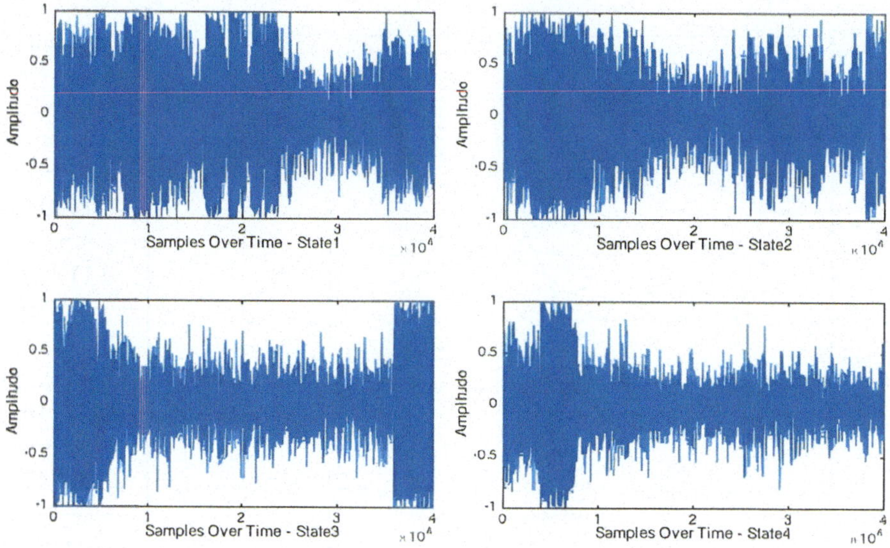

Fig. 4.4 Four different states of transition

equal-sized states. Here, an experiment was conducted where the input mix is considered as Iron ore, it takes around eight minutes to complete its crushing, with certain pre decided grinding parameters. So, there are four states decided here. The first state is marked as *State_Tran_*1, the beginning of the crushing. After two minutes of crushing, *State_Tran_*2 is marked while *State_Tran_*3 is represented as the state two minutes about to complete the crushing and the end of crushing is denoted by *State_Tran_*4. Four different transition states, as mentioned above, are shown in Fig. 4.4 and the corresponding amplitudes are stored as vectors for four states. For different raw materials, the crushing time is different, so the time gap between different states will be different.

4.2.2 Unique Pattern Generation

Density estimation is a statistical technique [13] used to infer the probability distribution of a random variable based on a finite sample of data. It involves estimating the underlying density function, representing the observed data. This process is essential in data analysis, pattern recognition, and machine learning, where understanding the data distribution is critical for decision-making and modelling. Different types of density estimation techniques are available to handle different data distributions, which are given as follows:

- Parametric Density Estimation—This technique assumes that the data follows a specific distribution (e.g., Gaussian, Exponential etc.). The parameters of the

distribution (mean, variance, etc.) are estimated from the data using methods like Maximum Likelihood Estimation (MLE), Method of Moments etc. This technique is highly dependent on the correctness of the assumed model.
- Non-Parametric Density Estimation—This technique does not assume a specific form for the underlying distribution. It is flexible and adaptable to the data distribution. Procedures used for this technique are Histogram, Kernel Density Estimation (KDE), and Nearest-Neighbour Methods.

Kernel Density estimator (KDE) [1, 14, 15] is applied here to predict each state of the running mill. Kernel estimators are used to smooth out the signal, considering the contribution of each observed data point of the signal based on its local neighbourhood. The contribution of the ith data point, say $x(i)$ estimates the point x^*, which indicates how apart they are. The shape of the contribution depends on the kernel function adopted with variable width (bandwidth). If the kernel function is denoted as K and the bandwidth as h, then the estimated density at any point x can be defined using Eq. (4.1).

$$\hat{f}(x) = \frac{1}{n}\sum_{i=1}^{n} K\left(\frac{x - x(i)}{h}\right) \tag{4.1}$$

where n number of data points are considered and $\int K(t)dt = 1$, to ensure that the estimated function $f(x)$ integrates all the data points within the transition time frame representing the unit area. The kernel function K is usually chosen to be a smooth unimodal function with a peak at zero. The most used kernel function is the Gaussian Kernel function, defined in Eq. (4.2), where x represents any given data point.

$$k(x) = \frac{1}{\sqrt{2\pi}} \exp\left(-\frac{1}{2}x^2\right) \tag{4.2}$$

The quality of the estimated signal does not depend much on the shape of K, but on the value of its bandwidth, i.e. h. It is important to choose the most appropriate value of the bandwidth, which is not too small or too large. The small value of h leads to a spiky estimated signal, while the large value leads to over-smoothing, as shown in Fig. 4.5a and b, respectively for a given dataset.

The transition segments are analysed to extract the knowledge for different states of grinding. Here, the estimation of the signal of the segmented data for four assigned states discussed in Sect. 4.2.1, with bandwidth h is assumed to be 100. After applying KDE, the patterns are generated for the segments in the load-varying conditions of the ball mill. For different raw materials, the same procedure can be applied to generate the unique patterns and store the corresponding information within the knowledge base. Also, estimation is generated for empty mills, i.e. without any load condition of the mill. Figure 4.6a and b represent the kernel estimation of a loaded and an empty ball mill with Iron ore as input. It is noticed that due to the presence of particles, the

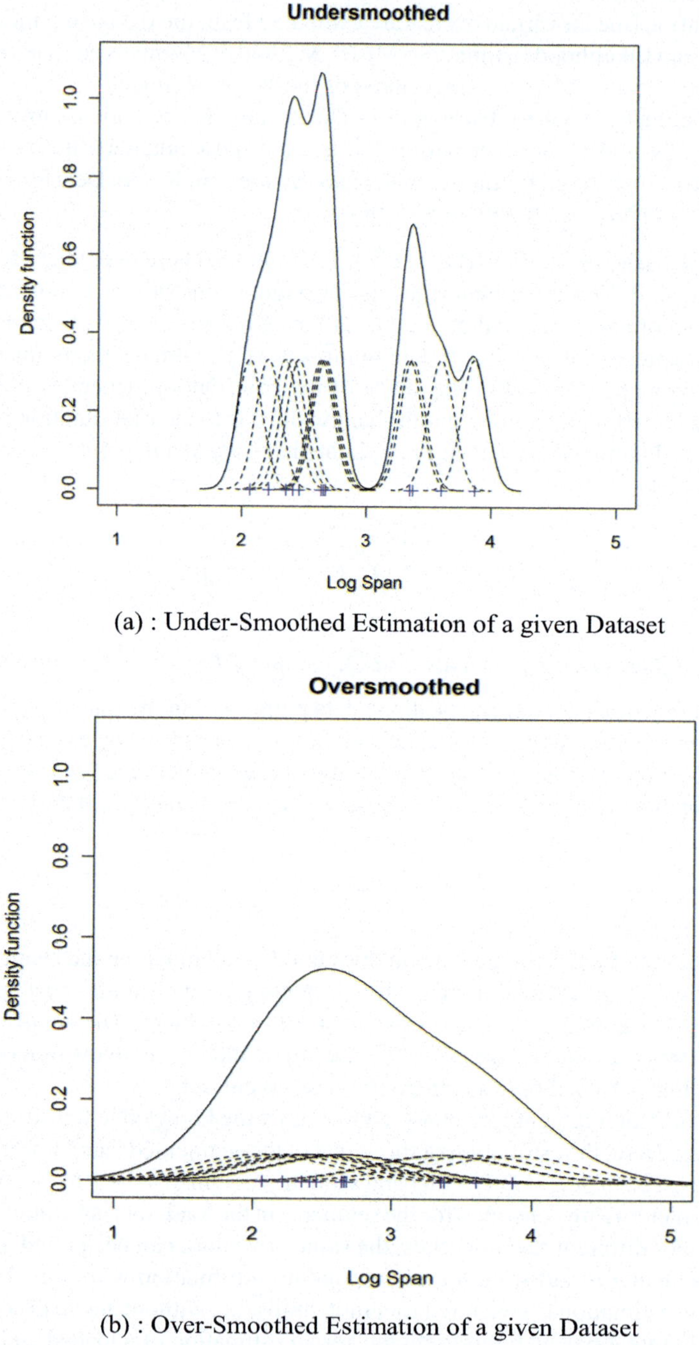

Fig. 4.5 **a** Under-smoothed estimation of a given dataset. **b** Over-smoothed estimation of a given dataset

peak value of the estimated density of the loaded mill is near to one, whereas for the empty mill, it is 1.5, which helps to differentiate the patterns.

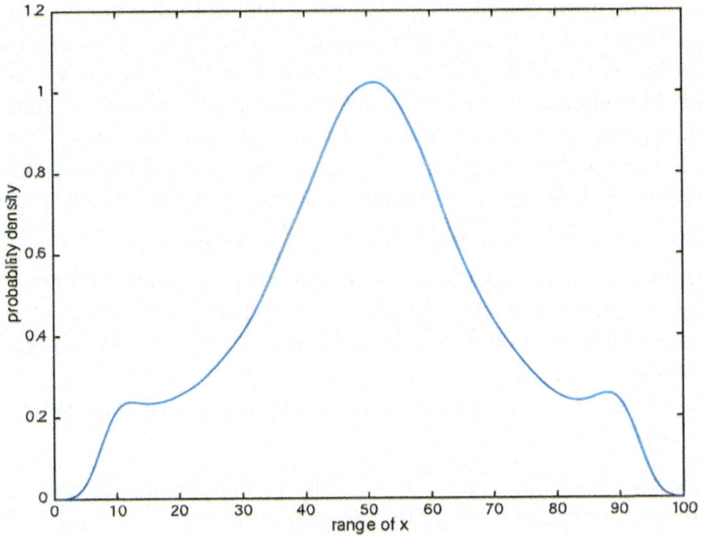

(a) : KDE of Loaded Ball Mill

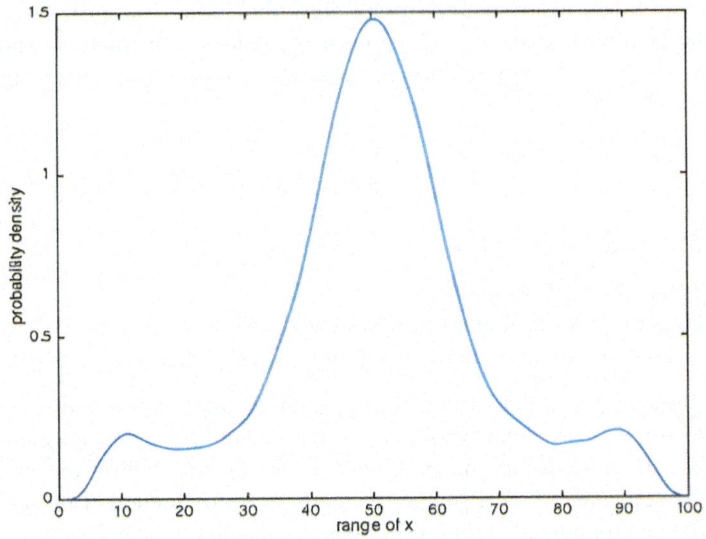

(b) : KDE of an Empty Ball Mill

Fig. 4.6 **a** KDE of loaded ball mill. **b** KDE of an empty ball mill

4.2.3 Curve Fitting

Curve fitting is the process of finding a mathematical function that best fits a set of data points. It involves determining a curve (or function) that closely approximates the relationship between variables in the data. Curve fitting is widely used in data analysis, predictive modelling, engineering, and scientific research. The main objective of Curve Fitting is to provide a compact and interpretable representation of data. This enables predictions for values outside the observed data range (extrapolation) or within the data range (interpolation). Ensure the curve minimises the difference between observed data and the predicted or simulated values. The following are the different types of curve fitting procedures.

- Linear Fitting assumes a straight-line relationship between variables.
- Non-linear fitting is used when data relationships are not linear.
- Parametric fitting assumes the form of the function is known (e.g., linear, polynomial).
- Non-parametric does not assume a specific functional form (e.g., splines or kernel regression).

For a specific type of input mix, the patterns are generated after applying KDE on different transition segments of the mill acoustics. During the operation, it has been noticed that the patterns are unique for each different state, as mentioned in Sect. 4.2.1. The shape of the estimated pattern is approximated as a bell-shaped function, which has been fine tuned using a curve-fitting algorithm. Since the distribution is bell-shaped, the Gaussian curve fitting technique is used to fit the patterns and generate the model. The curve fitting function is defined in Eq. (4.3) for a given signal x:

$$f(x) = \sum_{i=1}^{n} a_i e^{\left(\frac{-(x-b_i)}{c_i}\right)} \tag{4.3}$$

where a, b and c are three real-valued parameters,

a represents the peak value,
b is the position of the center of the peak and.
c is the standard deviation that controls the width of the curve.

The estimated signal is then tuned to generate Gaussian distribution by varying the parameters of the model. In this case, the parameter 'b' is fixed since the centre of the peak is at zero, while parameters 'a' and 'c' are tuned to follow the experimental results. The segments obtained experimentally are thus modelled using Eq. (4.3), and at each state for a fixed input mix, values of parameters (a, b and c) are obtained. Such information is stored in a look up table for different segments with respect to different raw materials.

4.2.4 Error Measurement

Unique patterns are now generated for different states of the running ball mill and fine tuned using Gaussian curve fitting algorithm. The experimental results always do not follow the simulated patterns, so, similarity between the two signals–(i) simulated and (ii) experimental is calculated using Euclidean distance given in Eq. (4.4).

$$d = \sqrt{\sum_{i=1}^{p} (x_i - y_i)^2} \tag{4.4}$$

where, x_i is the ith sample point obtained by experimentation and estimated using KDE while y_i is the simulated output of the same point and p is the total number of sample points present within the input segment. For different states of a given input mix, the error is calculated and stored in the knowledge base.

4.2.5 Simulated Patterns

Before interpreting any results, it is essential to consider the context, assumptions, and limitations of the methods used. The data quality, preprocessing steps, and model choice play a significant role in shaping the outcomes, and potential biases or noise

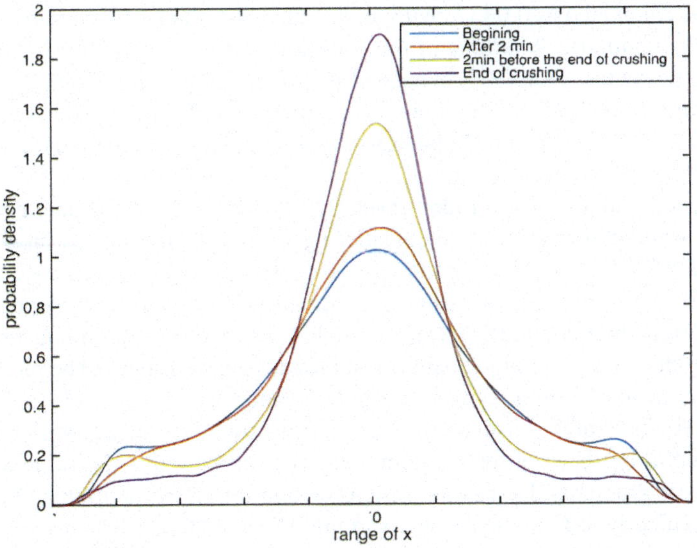

Fig. 4.7 Kernel density estimator for coal for four different states

in the data can influence the reliability of the findings. Additionally, understanding the evaluation metrics and their appropriateness for the specific task ensures that the results are meaningful. It is equally important to account for any randomness or variability, especially in experiments involving stochastic processes or small datasets. Clear documentation of these factors allows for informed discussions, ensuring that conclusions drawn are robust and grounded in a thorough understanding of the underlying processes. Here, the experiment is performed using different types of raw materials. The experimental results are shown for coal ore as the input raw material. Generated patterns using KDE for different load-varying conditions of the mill are shown in Fig. 4.7, distinguishing four different states of the running ball mill in its load-varying conditions. The peak values of KDE for *State_Tran_*1 and *State_Tran_2*, representing the beginning of crushing and two minutes after the crushing, respectively, are very close to each other. Since in both states, the size of the raw materials is larger, the impact between the material and media is very high, which produces high sound intensity. However, the peak value is lower. Gradually, with the crushing, the impact is reduced, so the peak value becomes higher. In comparison to *State_Tran_3* (2 min before the end of crushing), the peak value of *State_Tran_4* (End of crushing) is high. In Fig. 4.8, pattern representation of Iron has been demonstrated. The output of the Gaussian curve fitting algorithm, considering input as coal is shown in Fig. 4.9. It has been observed from the Figure that the simulated pattern follows the experimental pattern, though not exactly the same. The experimental and the simulated outcomes are sampled at different points and corresponding probability density values are framed in two vectors. Euclidean distance between the two vectors is calculated to examine that they are approximately equal. From Fig. 4.9, the experimental and simulated density values are obtained using which Euclidean distance is calculated as 0.0432. With all different raw materials the same procedure is performed and after fine tune of the parameters (Eq. 4.3), the error is measured, which is always less than 0.05, in each case. So it can be claimed that the simulated output is approximately 95% similar to the experimental results. Parameters are also within 95% of the confidence boundary, which implies the simulated output follows the experimental results.

Tables 4.2 and 4.3 give an idea about the tuned values of the parameters a and c for the simulated output while considering Coal and Iron ore as input material, respectively. The value of the parameter b is near to zero in each case because it represents the centre of the curve and according to KDE, the peak is always achieved at zero. From the knowledge base, it is noted that the corresponding *psd* of each state correlates with the unique patterns of the four derived states. The plotting of the parameters (a and c) with respect to grinding time for raw material Coal, is shown in Fig. 4.10. This plot considers all the different states from beginning till the end of crushing. It has been observed from Tables 4.2 and 4.3 that the value of standard deviation c is gradually decreasing, which corroborates the functioning of a ball mill. More specifically, c is higher at the beginning of crushing when particle sizes are large and gradually reaches the minimum value at the end of crushing. The parameter a is gradually increasing with the progress of crushing.

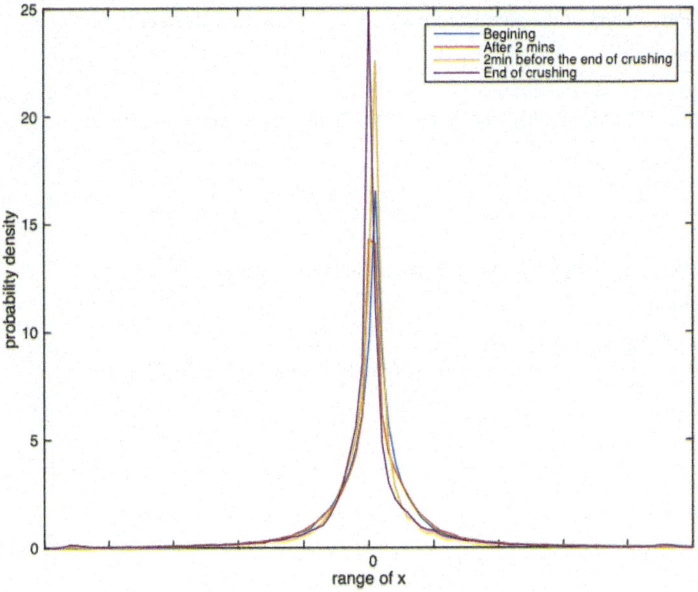

Fig. 4.8 Kernel density estimator for iron for four different states

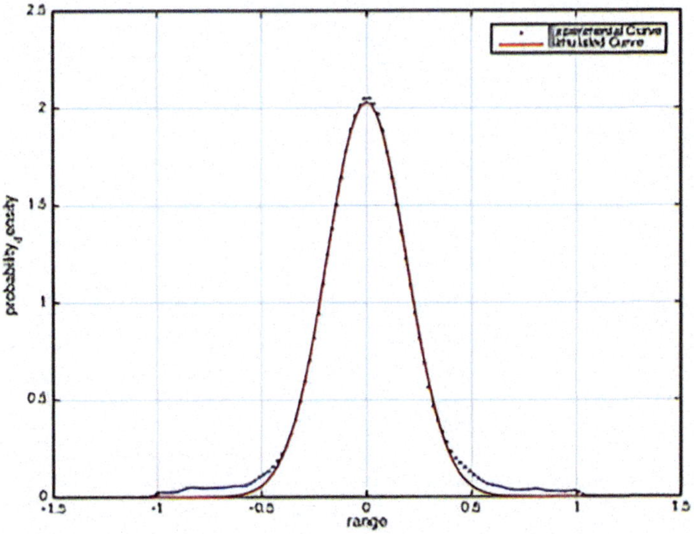

Fig. 4.9 Simulated versus experimental output for *State_Tran_3* of iron ore as input

Table 4.2 Parameters of the simulated patterns for coal at four different states

Load condition	*a*	*c*
Different load varying conditions	0.9227	0.5976
INPUT MIX—COAL includes the states of the beginning of the crushing till end	1.504	0.3179
	1.821	0.2779
	1.939	0.2666

Table 4.3 Parameters of the simulated patterns for iron at four different states

Load condition	*a*	*c*
Different load varying conditions	9.296	0.04991
INPUT MIX—IRON includes the states of the beginning of the crushing till end	11.37	0.04099
	15.73	0.02027
	20.62	0.0156

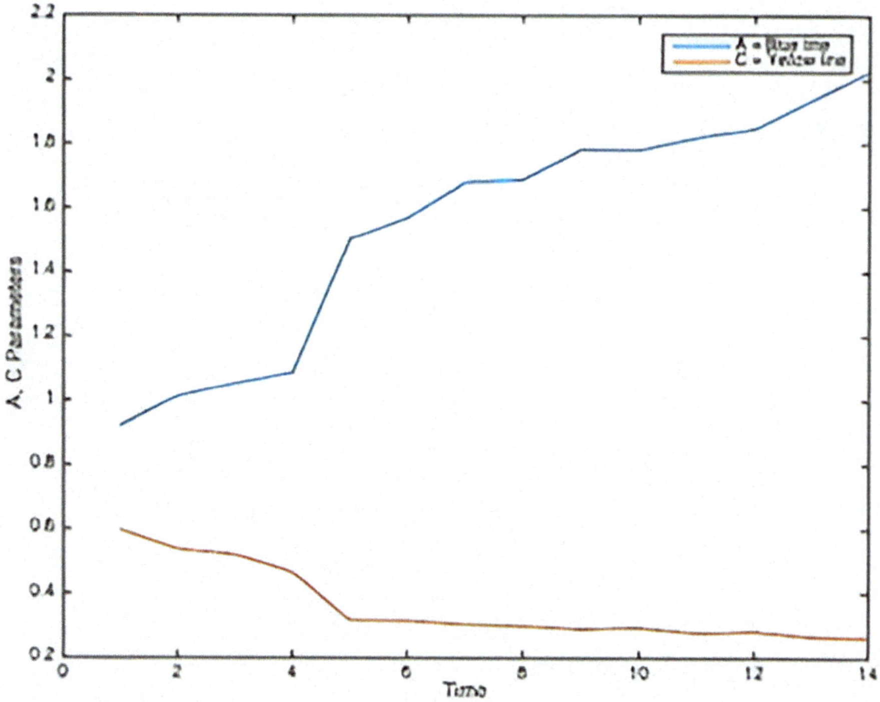

Fig. 4.10 Parameters *a* and *c* with respect to time for coal

It has been observed from the acoustic signatures that they are different in various load-varying conditions. Therefore, the proposed approach is successfully verified to predict the running states of the ball mill.

4.3 Feature-Based State Prediction

The second method predicts the terminating state by analysing the features extracted from the captured acoustics. A ball mill, while starting its operation with an input mix and a known set of grinding media, can be stopped with some prior criteria when the expected desired particle size distribution (*psd*) has been achieved. To predict the terminating state, information related to that state needs to be identified. Generally, from an acoustics signal the three main features—amplitude, phase and frequency are measured, however these features are not enough for a prediction model. More features are extracted from the acoustics signal like—statistical moments, following works [16–18]. Principal Component analysis-based feature extraction method [19–22] is used to identify the meaningful features from the extracted features. Next the feature vectors of the dataset are clustered using the K-means algorithm where K represents Start state and End state. Finally, Naive Bayes classification algorithm is applied to identify the terminating state. The flow diagram of the proposed method is given in Fig. 4.11, and the detailed discussion is given in the subsequent subsections.

4.3.1 Preprocessing of Method 2

In this section, one full cycle of the mill vessel of the running ball mill is considered as an input segment and stored in a database, named Full_cycle. Full cycle of a mill means one complete cycle of the mill vessel, which contains the transition states. Figure 4.12 shows the representation of such a transition state, marked with a red box. The aim of this approach is to predict the end state of the grinding of a given input material. The segments of the start and end states are shown in Fig. 4.13. At the beginning, when the material is uncrushed, the state is marked as *Start State* and at the end of crushing i.e., when the required fine is generated, marked as *End State*.

To distinguish the Start and the End State of the running ball mill, the following steps are applied to the input acoustics. The full cycle segments are taken as input and converted into the time domain data, as discussed earlier, and apply the following steps to extract the features:

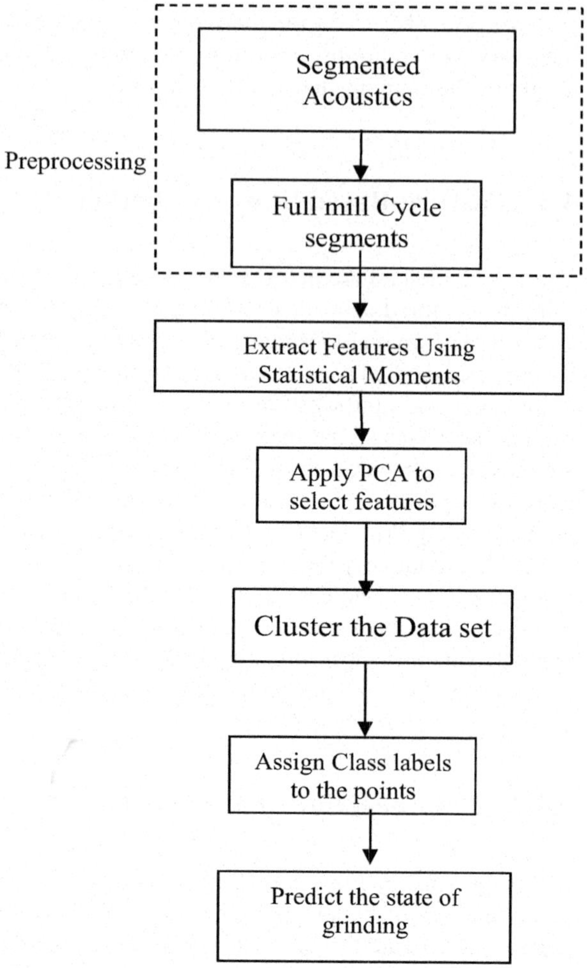

Fig. 4.11 Flow diagram of the feature-based end state prediction

Step 1: Statistical moments are extracted from each segment to build the feature vector for each full cycle and obtain an information matrix where segments and features represent the rows and columns, respectively.
Step 2: PCA is applied to the information matrix to find the principal components as the important features.
Step 3: Cluster the input segment using the k-means clustering algorithm to distinguish the Start and End state.
Step 4: Classify the data set to predict the two states (*Start State, End State*) of grinding.
Step 5: Performance evaluation is done for the model.

Following is a detailed discussion of each step.

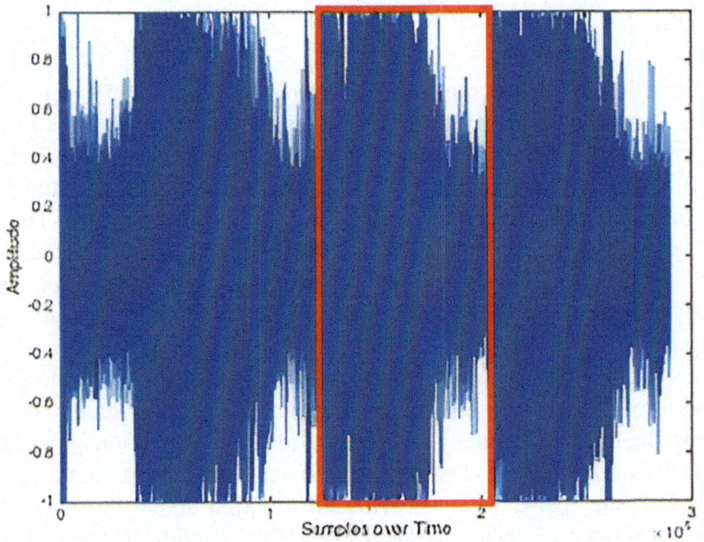

Fig. 4.12 One full cycle of a running ball mill

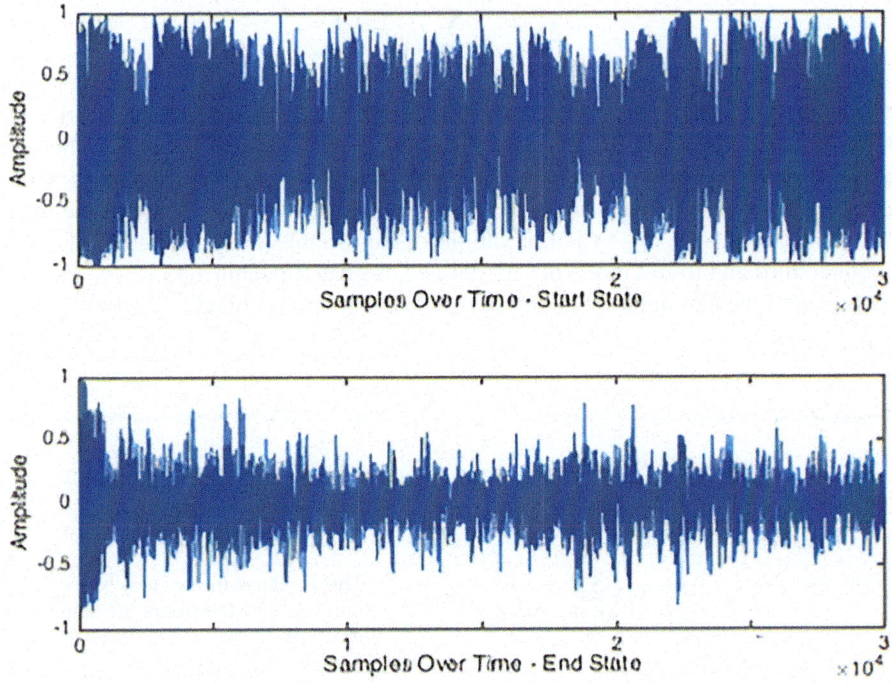

Fig. 4.13 Acoustic signals of *Start State* and *End State*

4.3.2 Feature Extraction

Feature extraction is a process in machine learning and data analysis aimed at transforming raw data into a set of meaningful and representative attributes, or features. By selecting the features, the dimensionality of the data is reduced, improving the computational efficiency and enables better model performance. Effective feature extraction involves identifying characteristics that are most relevant to the task. The quality of features significantly impacts the accuracy and interpretability of predictive models, making feature extraction a cornerstone of successful data-driven applications.

To analyse an acoustic signal, the common features that are generally considered are the amplitude, frequency, and phase. Statistical moments [23] are considered as other important features, integrated with the common features, and used for the classification of the running mill states whether start or end state. The mean and variance are the first two moments, which provide the information about the location and variability of the data set. The third and fourth statistical moments are skewness and kurtosis, respectively, which provide information about the shape of the data distribution. Using the general moment Eq. (4.5), different features are extracted, and the feature matrix is formed.

$$\pi_k = \frac{1}{N}\sum_{i=1}^{N}(x_i - \alpha)^k \tag{4.5}$$

where $x_1, x_2, \ldots .x_N$ are observations of a sample variable X, having its origin at some point α, (experimentally defined) and π_k the kth moment about α. When the actual mean is in fractions, moments are first calculated about an arbitrary point and then converted to moments about the actual mean. Now, the moments of the mean i.e., $\alpha = \overline{X}$ and $\left(s_i = x_i - \underline{X}\right)$ observations of the variable X, is considered. The first, second, third and fourth moments are defined as given in Table 4.4.

Standard Deviation based on the statistical moment is defined in Eq. (4.6).

$$\mathrm{S} = \sqrt{M_2} \tag{4.6}$$

Table 4.4 Statistical moments

First moment	$M_1 = \sum_{i=1}^{N} s_i = 0$	Represents the mean
Second moment	$M_2 = \frac{1}{N}\sum_{i=1}^{N} s_i^2 = M_2 - M_1^2$	Represents the variance
Third moment	$M_3 = \frac{1}{N}\sum_{i=1}^{N} s_i^3 =$ $M_3 - 3M_1M_2 + 2M_1^3$	This is related to skewness but is influenced by the size of the unit of measure
Fourth moment	$M_4 = \frac{1}{N}\sum_{i=1}^{N} s_i^4 =$ $M_4 - 4M_1M_3 + 6M_1^2M_2 - 3$	This is directly related to kurtosis but influenced by the size of the metric unit

Say Iron ore is taken as the input mix, and the experiment is conducted. The captured audio is fragmented, where 98 different samples are considered for denoting the start and end state of grinding of a given raw material to produce an expected *probability density* as an output. From those fragments, 10 (ten) features, as discussed earlier, are extracted. Thus, the information matrix of size 98×10 is constructed.

4.3.3 Principal Component Analysis

Principal Component Analysis (PCA) [24, 25] is a widely used dimensionality reduction technique. It transforms the high-dimensional data into a smaller set of uncorrelated variables called principal components while retaining as much variability as possible. PCA works by identifying the directions, or components, that maximize the variance in the dataset, calculated through eigenvectors and eigenvalues of the data's covariance matrix. These components serve as a new coordinate system for the data and prioritising the features that contribute the most to its variance. By reducing the number of dimensions, PCA improves computational efficiency, reduces noise, and helps visualise complex datasets in two or three dimensions. However, since PCA assumes linear relationships and focuses solely on variance, it might not capture intricate structures in the data, limiting its applicability in certain cases.

In this experiment, a covariance matrix is formed of size 10×10, which is symmetric in nature, generated from the information matrix, on which the PCA [26] is applied to find the most important feature. Then the eigenvector and eigenvalues of the covariance matrix are generated. From the eigenvector, a rank matrix has been formed and based on the rank matrix, the eigenvalues are considered as the Principal Components of the dataset. In this method, three of the positive eigenvalues according to rank are taken as the important components.

In general, PCA helps to reduce the number of 'features' while preserving the variance. On the other hand, clustering reduces the number of 'data points' by summarizing several points by their means (in the case of the k-means algorithm). So, if the dataset consists of d data points and f features each, then PCA claims to compress the f features, whereas clustering aims at compressing the d data points.

In this proposed method, the search space is then reduced by partitioning the data sets as—*Start State and End State.* After selecting features using PCA, k-means clustering algorithm is applied where k is 2, representing the *Start State* and the *End State.* Finally class labels are assigned to each of the data points using the Naive Bayes classifier and the information matrix is built based on the features (principal components) and the class labels.

4.3.4 Evaluation of the Model

Model evaluation is a crucial step in the machine learning workflow, aimed at assessing how well a model performs on unseen data. It involves using metrics and validation techniques to measure the accuracy, precision, recall, F1-score, or other performance indicators based on the problem type, such as classification, regression, or clustering. Cross-validation is a popular strategy, dividing data into training and testing subsets to ensure the model generalizes well. Techniques like confusion matrices, ROC curves, and mean squared error provide detailed insights into strengths and weaknesses. A thorough evaluation helps to identify the issues like overfitting or underfitting and informs decisions on model tuning, feature selection, or choice of algorithm. Proper evaluation ensures that the final model meets the intended objectives and performs reliability in real-world applications.

The training data set with class labels are used to model different classifiers and applied to classify the states of the mill using the test data set. There are indexes based on which the performance of the classifier is evaluated. The correctness of a classifier is evaluated by computing the number of correctly recognised class examples (*tp*: true positive), the number of correctly recognized examples that do not belong to the class (*tn*: true negative), the number of examples that are incorrectly assigned to the class (*fp*: false positive) and ultimately the number of examples that are recognized as class examples (*fn*: false negative). Table 4.5 shows the definition of performance measures [27] of classifiers.

The *Confusion Matrix* is a useful tool for analysing how well the classifier recognises the data set of different classes. The sensitivity and specificity measures are used to calculate the accuracy of the classifiers. ROC (Receiver Operating Characteristics) curves are plotted to visualize the performance of the multiclass classification problem. It provides information about how much the model is capable of distinguishing between classes. To divide the dataset into training and test sets, tenfold cross-validation process is applied, which validates the result.

Table 4.5 Performance measures

Measure	Formula	Evaluation focus
Accuracy	$\frac{tp+tn}{tp+tn+fn+fp}$	Overall effectiveness of a classifier
Precision	$\frac{tp}{tp+fp}$	Class agreement of the data labels with the positive labels given by the classifier
Recall (sensitivity)	$\frac{tp}{tp+fn}$	Effectiveness of a classifier to identify positive labels
Specificity	$\frac{tn}{fp+tn}$	How effectively a classifier identifies negative labels
AUC	$\frac{1}{2}\left(\frac{tp}{tp+fn}+\frac{tn}{tn+fp}\right)$	Classifier's ability to avoid false classification

Table 4.6 Confusion matrix

True positive (TP) (27)	False positive (FP) (16)
False negative (FN) (1)	True negative (TN) (54)

Table 4.7 Test mode: Naive Bayes classifier using tenfold cross-validation

Detailed accuracy by class	TP rate	FP rate	Precision	Recall	F-Measure
Class 1	0.628	0.018	0.964	0.628	0.761
Class 2	0.982	0.372	0.771	0.982	0.864
Weighted avg.	0.827	0.217	0.856	0.827	0.819

4.3.5 Simulated Outcome of Feature-Based Classification

Principal component analysis (PCA) method is applied on the experimental output represented as a 2-D matrix with size 98 $\times$ 11 where the 11th column represents the class labels with two values: *Start state* and *End state.* Naïve Bayes' classifier is applied, and the result in terms of the confusion matrix is given in Table 4.6. Other performance measures of the classifier output are shown in Table 4.7. The correctly classified instances of the experiment are approximately 83%. The receiver operating characteristic (ROC) curve area is similar to a life chart, used in signal detection to show the trade-off between hit rate and false alarm rate over noisy channels. The experiment provides the ROC area is 0.936 for all classes. The ROC graph is shown in Fig. 4.14.

4.4 Summary

In this Chapter two methods are discussed to simulate the different states of grinding and to predict the end state for a given input. A novel approach is discussed to identify the running states of the ball mill based on a mathematical model. Based on the transition cycles of the running ball mill, a unique pattern is generated for each state with a fixed input mix. Then the model is proposed to identify the four running states of the ball mill operation as discussed in Sect. 4.2.1. Finally, the pattern is reconstructed using a curve fitting algorithm. Depending on the shape of the pattern the curve fitting function is chosen. The proposed mathematical function is given in Eq. (4.3), which is a polynomial function with three parameters (*a*, *b*, and *c*) and fine tune them so that the simulated output must follow the experimental result. Ultimately, using Euclidean distance the error is minimized so that simulated output

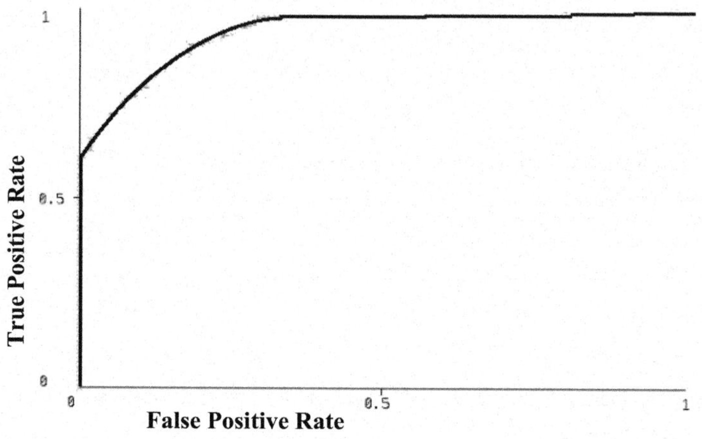

(a) : ROC for Class 1 - *Start State*

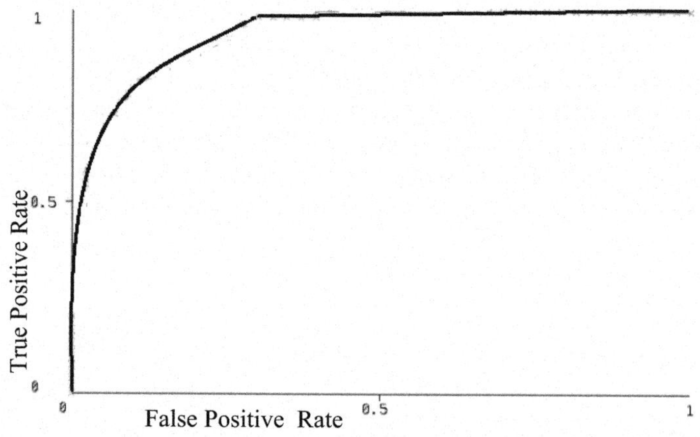

(b) : ROC for Class 2 – *End State*

Fig. 4.14 **a** ROC for class 1—*Start State*. **b** ROC for class 2—*End State*

can follow the experimental output. Under different training conditions convergence of error in acoustic signature has been investigated and obtained satisfactory results.

The second method predicts the End state of the running ball mill using the extracted statistical features obtained by analysing the mill acoustics. The results can predict the desired *probability density* of the input mix at different states of the running ball mill and correlate the same with the experimental results and can be deployed in the mining industry by fabricating a system on a chip.

Acoustic analysis of the ball mill is an integral part of the work. It is concluded that the mathematical and classification model are useful to predict different running states of the ball mill including End state by leveraging the information obtained from the size band of the raw material, stored as per experiment of different crushing materials. Its importance lies with the formation of the knowledge base of the proposed model and the generation of a feedback signal that is the key factor in a closed-loop information system. The model is not only valid for the ball mill but also expected to work well with other different kinds of reactors/crushers/compound machines. Under different training conditions convergence of error in acoustic signature has been investigated and obtained satisfactory results.

Exercise

1. What is the purpose of using Kernel Density Estimation (KDE) in processing Audio Signal?
2. How does KDE differ from Histogram in estimating probability density function?
3. Explain use of curve fitting algorithms for pattern recognition?
4. Which methods can be used to extract parameters from the Acoustics?
5. Where is the utilization of PCA in Parameter Gradation?
6. How the error in minimization between simulated and experimental results in reference to curve fitting?

References

1. Wahiduzzaman, M., Yeasmin, A.: A kernel density estimation approach of North Indian Ocean tropical cyclone formation and the association with convective available potential energy and equivalent potential temperature. Meteorol. Atmos. Phys. **132**, 603–612 (2020). https://doi.org/10.1007/s00703-019-00711-7
2. Lahodny Jr., G.: Curve Fitting and Parameter Estimation, pp. 1–18 (2015)
3. Gurley: Curve fitting techniques. In: Numerical Methods Lecture 5, CGN 3421—Computer Methods, pp. 86–99 (2018)
4. Huang, C.W., Lee, K.C.: Application of ICA technique to PCA based radar target recognition. Prog. Electromagn. Res. **105**, 157–170 (2010)
5. Hoppe, E., Roan, M.: Principal component analysis for emergent acoustic signal detection with supporting simulation results. J. Acoust. Soc. Am. **130**(4), 1962–1973 (2011). https://doi.org/10.1121/1.3628324
6. Jiang, Y., Jin, X.: Using k-means clustering to classify protest songs based on conceptual and descriptive audio features. In: Rauterberg, M. (ed.) Culture and Computing. HCII 2022. Lecture Notes in Computer Science, vol. 13324. Springer, Cham (2022). https://doi.org/10.1007/978-3-031-05434-1_19
7. Smola, A., Vishwanathan, S.V.N.: Introduction to Machine Learning. Cambridge University Press, pp. 33–36 (2008)

8. Pryadko, N.: Optimization of Fine Grinding on the Acoustic Monitoring Basis, Energy Efficiency Improvement of Geotechnical Systems, pp. 101–108. Pivnyak, Beshta & Alekseyev, Taylor & Francis Group, London (2014). ISBN 978-1-138-00126-8
9. Desobry, F., Davy, M., Fitzgerald, W.J.: Density kernels on unordered sets for kernel-based signal processing. In: 2007 IEEE International Conference on Acoustics, Speech and Signal Processing-ICASSP'07, vol. 2, pp. 417–421. IEEE (2007). https://doi.org/10.1109/ICASSP.2007.366261
10. Baranowski, J., Piątek, P., Kawala-Janik, A., Pelc, M., Anthony, R.J.: Application of Kernel density estimators for analysis of EEG signals (2012)
11. Zheng, Y., Jestes, J., Phillips, J.M., Li, F.: Quality and efficiency for kernel density estimates in large data. In: Proceedings of the 2013 ACM SIGMOD International Conference on Management of Data, pp. 433–444 (2013). Copyright 2013 ACM 978-1-4503-2037-5/13/06
12. Danese, M., Lazzari, M., Murgante, B.: Kernel density estimation methods for a geostatistical approach in seismic risk analysis: the case study of Potenza Hilltop Town (Southern Italy). In: Gervasi, O., Murgante, B., Lagan, A., Taniar, D., Mun, Y., Gavrilova, M.L. (eds.) Computational Science and Its Applications—ICCSA 2008. ICCSA 2008. Lecture Notes in Computer Science, vol. 5072. Springer, Berlin, Heidelberg (2008). https://doi.org/10.1007/978-3-540-69839-5_31
13. Choi, M., Sweetman, B.: Efficient calculation of statistical moments for structural health monitoring. Struct. Health Monit. **9**(1), 13–24 (2010)
14. Yu, L., Su, Z.: Application of kernel density estimation in lamb wave-based damage detection. Math. Probl. Eng. **2012**(1), 406521 (2012). https://doi.org/10.1155/2012/406521
15. Zucchini, W.: Applied Smoothing Techniques Part 1: Kernel Density Estimation, pp. 1–20 (2003)
16. Soliman, S., Hsue, S.-Z.: Signal classification using statistical moments. IEEE Trans. Commun. **40**, 908–916 (1992). https://doi.org/10.1109/26.141456
17. Lopatka, J., Pedzisz, M.: Automatic modulation classification using statistical moments and a fuzzy classifier. In: WCC 2000-ICSP 2000. 2000 5th International Conference on Signal Processing Proceedings. 16th World Computer Congress 2000, vol. 3, pp. 1500–1506. IEEE (2000). https://doi.org/10.1109/ICOSP.2000.893385
18. Suciati, N. et al.: Feature extraction using statistical moments of wavelet transform for iris recognition. In: 2016 International Conference on Information & Communication Technology and Systems (ICTS), pp. 193–198 (2016)
19. Castells, F., Laguna, P., Srnmo, L., et al.: Principal component analysis in ECG signal processing. EURASIP J. Adv. Signal Process **2007**, 074580 (2007). https://doi.org/10.1155/2007/74580
20. Palma, L.B., Coito, F.V., Gil, P.S., Neves-Silva, R.: Process control based on PCA models. In: 2010 IEEE 15th Conference on Emerging Technologies and Factory Automation, ETFA 2010, pp. 1–4. IEEE (2010). https://doi.org/10.1109/ETFA.2010.5641080
21. Crida, R., Stoddart, A., Illingworth, J.: Using PCA to model shape for process control, pp. 318–325 (1997). https://doi.org/10.1109/IM.1997.603882
22. Moore, B.: Principal component analysis in linear systems: Controllability, observability, and model reduction. IEEE Trans. Autom. Control **26**(1), 17–32 (1981). https://doi.org/10.1109/TAC.1981.1102568
23. Boveiri, H.R.: On pattern classification using statistical moments. Int. J. Signal Process. Image Process. Pattern Recognit. **3**(4), 15–23 (2010)
24. Gupta, V., Mittal, M.: Respiratory signal analysis using PCA, FFT and ARTFA. In: 2016 International Conference on Electrical Power and Energy Systems (ICEPES), pp. 221–225 (2016). https://doi.org/10.1109/ICEPES.2016.7915934

25. Mishra, A.K., Mulgrew, B.: Radar signal classification using PCA-based features. In: 2006 IEEE International Conference on Acoustics Speech and Signal Processing Proceedings, p. III (2006). https://doi.org/10.1109/ICASSP.2006
26. Hamadache, M., Lee, D.: Principal component analysis based signal-to-noise ratio improvement for inchoate faulty signals: application to ball bearing fault detection. Int. J. Control Autom. Syst. **15**(2), 506–517 (2017). https://doi.org/10.1007/s12555-015-0196-7
27. Sokolova, M., Lapalme, G.: A systematic analysis of performance measures for classification tasks. Inf. Process. Manage. **45**(4), 427–437 (2009)

Chapter 5
Residence Time Calculation for Grinding the Materials Using Genetic Algorithm

Abstract This Chapter discusses how to determine the residence time of breakage of the materials in the ball mill in order to forecast any state of grinding of the ball mill. Using Genetic Algorithm the desired state of grinding of the ball mill is predicted based on its two previous consecutive states and after running the ball mill up to that predicted state, it is stopped automatically. At that desired state, the mill comes to an end, and the targeted particle size distribution (*psd*) of the material is collected. Discussion follows how the optimum parameters of grinding are decided to stop the mill by designing a fitness function, which minimizes material wastage. Next the discussion is on the procedure to correlate the particle size distribution at different states of grinding and determining the residual time to stop the mill when the desired breakage particles are acquired, thus minimizing the loss of materials. The particle size distribution of the materials at different states of grinding are obtained by applying Gaussian Mixture Model (GMM) and clustered using fuzzy c-means (FCM) algorithm where number of groups are equal to the predetermined states of the ball mill. When the mill stops, the *psd* of the breakage material corresponding to a particular state must belong to the respective cluster, verifying the experimental result.

5.1 Introduction

In the mining industry, waste of raw materials is the main concern. Raw material is crushed within the ball mill as a closed loop manner and the operation is maintained manually [1–3]. The system is operated based on the operator's experience, like when to stop the crusher for checking with a particular type of input raw material. Due to the involvement of humans, there is a chance of over-crushing the material, which leads to the wastage of valuable minerals. So, in this closed-loop manual grinding method, after each unit of residence time, the mill needs to be stopped manually, and then the material is screened using a sieve shaker to check whether the desired size distribution has been achieved or not [4, 5]. This method is continued iteratively, though nobody can give a guarantee regarding the full utilization of the materials

J. Sil et al., *Intelligent Real-Time Control Systems to Predict the Particle Size of Minerals Using Acoustics of Ball Mill*, Studies in Systems, Decision and Control 600,
https://doi.org/10.1007/978-3-031-91290-0_5

due to the chance of over-crushing. Over-crushing in the ball mill occurs when the grinding process continues beyond the desired particle size, leading to the production of excessive fines or powder. This phenomenon results in wastage of raw material as the ultra-fine particles generated are often not suitable for the intended application and may require additional processing or be discarded entirely. Several factors are involved in over-crushing of material that may include incorrect mill settings, prolonged grinding time, or inappropriate feed size. When raw materials are crushed too finely, their surface area increases disproportionately, potentially altering their physical and chemical properties. This can reduce the material's usability in subsequent processes, such as blending, compaction, or sintering, where a specific particle size distribution is critical. Moreover, over-crushing leads to unnecessary energy consumption, increased wear and tear on the mill components, and higher operational costs. To mitigate this issue, operators should monitor grinding parameters closely, use appropriate liners and grinding media, and establish real-time process controls to halt grinding, once the desired particle size is achieved.

As discussed in Chap. 2, the continuous sound of the running ball mill in load-varying conditions is captured and the signal is segmented based on the one full cycle of the vessel or the transition cycles of the mill, shown in Fig. 5.1a and b, respectively. This continuous signal is captured for the total time period of crushing of a particular material from the beginning till end of crushing and that is divided into a number of segments based on the criteria of one full mill vessel cycle or based on the transition of the vessel from one cycle to the next. The segments are represented as the states of running, out of which few important segments are considered for state prediction of the running ball mill. During crushing operation of the ball mill with load, the total residence time (start state to end state) has been divided into four equal time intervals and each of them are represented as one of the States. At the end of each interval the mill is stopped manually in order to obtain information about the *psd* of corresponding breakage material. Each of the time intervals is considered as a segment which are equidistant and represented as State1, State2, State3 and State4. Each of the segments in their time domain representations are also shown in Fig. 5.2. The time domain data are sampled representing amplitudes and stored as an 1-D array.

It has been verified that at different states of grinding, particle size distribution (*psd*) has been estimated analytically which is close to the experimental results of corresponding states. Based on the relation between the states of grinding and *psd* of materials, this chapter addresses the issue of over-crushing by achieving desired *psd* at a particular state with an aim to minimise the wastage of materials. The proposed method predicts after how many full rotations of the mill vessel (full cycle), the system needs to be stopped. The residence time, defined as the residencel time duration of the material in the mill, is predicted by calculating the particular state of grinding from two of its previous consecutive states using Genetic Algorithm (GA) [6, 7] and at that state, the mill is stopped automatically. At least information of the first two segments of the running ball mill is required in order to calculate the third state. In

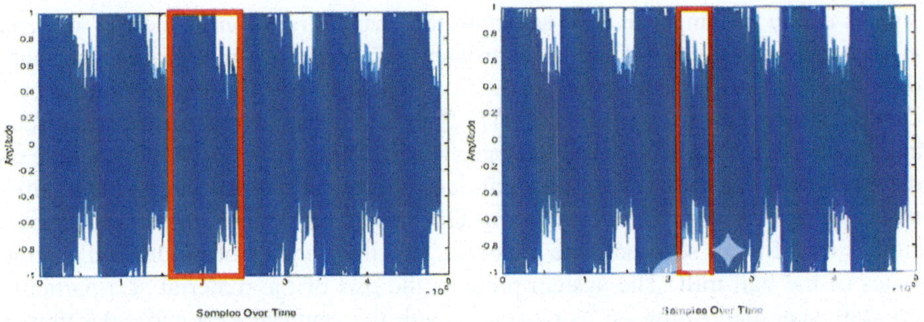

a) Continuous Acosutics with red coloured segment as full cycle

b) Continuous Acosutics with red coloured segment as transition cycle

Fig. 5.1 Segments from the continuous acoustics

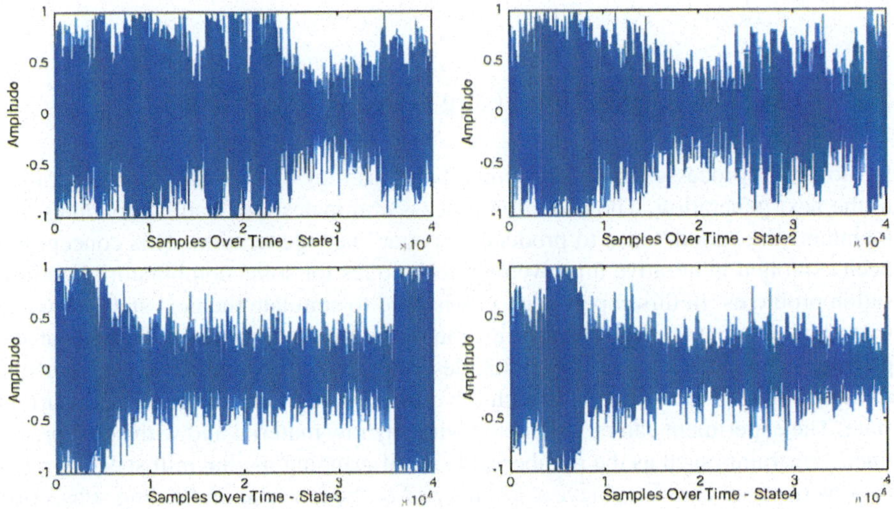

Fig. 5.2 Time domain representation of four different states

general, information of segments $(S(t), S(t+1))$ at t^{th} and $(t+1)^{th}$ time instants are considered to obtain the information of segments at $(t+2)^{th}$ time instant. Thus the closed loop real-time grinding process has been simulated using GA. The method calculates the residence time of crushing from the current state and accordingly terminates the mill, achieving desired *psd* of the material.

Next Gaussian Mixture Model (GMM) [8] is invoked to obtain the distribution of the particle size at different states of grinding and by applying Fuzzy c-Means (FCM) [9, 10] clustering algorithm the particles are grouped corresponding to the number of states of grinding of the ball mill. GMM [11] is applied to the segmented signals of the running ball mill to obtain the size distribution of the raw material at each state of crushing. GMM is a probabilistic model for representing normally distributed sub-populations based on the overall population. The distribution has been clustered using Fuzzy C-means (FCM) [12] algorithm, where each cluster represents a particular size range of the raw material. The FCM algorithm is chosen because the distributions at different states are continuous and may overlap due to the additive nature of the signals. Therefore, using GMM and FCM algorithms, the acoustic signals of different segments are clustered into the number of predefined running states of the ball mill. The system predicts the *psd* of the material at a particular grinding state depending on its closeness with the centroid of a particular cluster. The simulated outputs are verified with the experimentally screened output showing equivalent distribution patterns at any particular state. The proposed method predicts the residence time, mentioning the number of rotations of full cycles of the mill vessel required before stopping the mill. After the ball mill stops, the targeted particle size has been achieved which avoids over-crushing of the material and minimises wastage of material.

5.2 Next State Prediction Using Genetic Algorithm

In biological processes, species become better as they evolve from one generation to the next generation. The evolution process eliminates the "bad" individuals and maintains the "good" ones to produce a "better" new generation. This concept has been exploited in iterative improvement techniques for some combinatorial optimisation problems. In this experiment, GA is used to simulate the next state from any two previous consecutive states of a running ball mill. The block diagram of the proposed method is shown in Fig. 5.3, describing how the number of cycles of the running ball mill are utilized to reach the desired state of crushing from the current state. The experiment can be conducted with any raw material and with some predefined constraints such as the number and size of grinding media, mill speed, type of the raw material etc., to achieve a specific *psd* of the material. Following subsection contains a detailed discussion on the method.

5.2.1 *Initialization of Population*

Genetic Algorithm (GA) is a population based stochastic search algorithm where each individual in the population is considered as a candidate solution. In the proposed algorithm, amplitudes at different sampling points of time domain signal are represented using binary encoded chromosomes of length eight. The size of the initial population is 30 k, i.e. 30 k number of chromosomes.

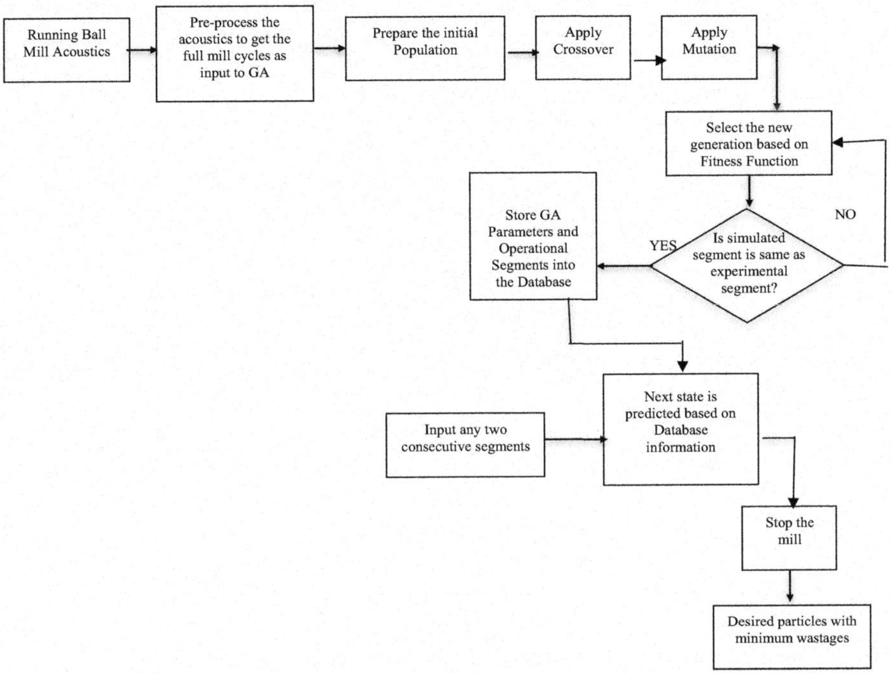

Fig. 5.3 Flow diagram of the GA-based prediction method

Before representing the amplitude into binary form, following steps are executed.

Step 1: Signed amplitude is represented as below:

- 1 bit for the sign (0 for positive, 1 for negative).
- 7 bits for the magnitude (value).

Step 2: Scale the value to determine the Range:

- The values are scaled within 0 to 127

Step 3: Convert the amplitude value into signed magnitude representation using 8 bit Chromosome.

For example, if the absolute value of a given amplitude is – 0.117240.

Then Scale the value to fit within the range [0–127] i.e. $0.11724 \times 127 = 14.893$, which is round of to the nearest integer i.e. 15.

Then the integer representation of the amplitude using 7 bit binary is 0001111.

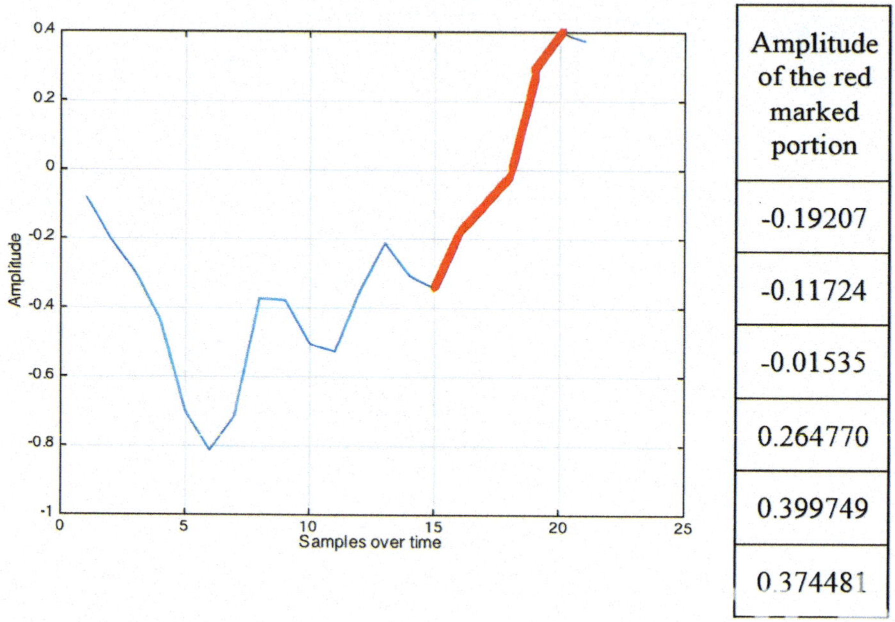

Fig. 5.4 Amplitude array for the signal highlighted

Table 5.1 Amplitude versus encoded value

Amplitude	Encoded value
− 0.19207	10011000
− 0.11724	10001111
− 0.01535	1000 0010
0.264770	0010 0010
0.399749	0011 0011
0.374481	0011 0000

Finally, the Sign bit is added to obtain the amplitude representation in binary: 10001111.

For example, the amplitude array of Fig. 5.4 is encoded as binary chromosomes, given in Table 5.1.

5.2.2 *Selection of Population*

After initialization of population, the next stage is selection of populations or chromosomes. The goal of the population selection mechanism is to ensure that better-performing individuals have a higher chance of reproduction in the new generation.

Population selection balances the exploitation of the best solutions with an exploration of the search space, ensuring diversity and avoiding premature convergence to suboptimal solutions. In the proposed algorithm, from the initial population, each pair of individuals are selected, represented as chromosomes of two consecutive states of the running ball mill and used as the pair of parents for crossover operation. Information of segments $(S(t), S(t+1))$ at t^{th} and $(t+1)^{th}$ time instants are selected as pairs of parent chromosomes for crossover operation where t is a positive integer.

5.2.3 *Crossover Point*

Crossover is a fundamental genetic operator, used to combine the genetic information of two individuals as parent chromosomes in order to generate offspring, mapped as new individuals in the population of the next generation. The crossover operation involves selecting one or more crossover points along the chromosome, which represent the locations where the parent's genetic material is exchanged. The choice of crossover points is crucial as it influences the diversity of offspring and the algorithm's ability to explore the solution space effectively. Here, the one-point crossover operation [13] is applied, where crossover point is calculated using Eq. (5.1). The point divides the chromosome into two segments and one segment from each parent is swapped to form the offspring.

$$p = (k + i) \tag{5.1}$$

Crossover point p is obtained by adding bit position k $(0 \le k \le 6)$ of the selected pair of chromosome with i $(1 \le i \le 7)$ so that p *is always* less than 8, i.e. length of the chromosome. The offspring are generated from each pair of chromosomes, and they are evaluated using the fitness function in order to take part in the next generation.

5.2.4 *Mutation*

Mutation in Genetic Algorithms (GAs) is a critical operator that introduces random changes to an individual's genetic representation to maintain diversity in the population and explore new areas of the solution space. By altering one or more gene values in a chromosome, mutation helps to prevent the algorithm from stagnating in local optima.

5.2.5 Fitness Function

Before calculating fitness value of each offspring chromosome, which is binary coded, convert that into decimal value, representing corresponding amplitude value. The fitness function [14] $f(i)$ is defined in Eq. (5.2),

$$f(i) = \frac{1}{e^{-\alpha x(i)}} \tag{5.2}$$

where α is a constant and $x(i)$ is the variable representing the amplitude of i-th offspring. Appropriate value of $\alpha = 0.57$ has been derived through simulation, considering minimum Euclidean distance between the simulated and the experimental results at the same sample point of a particular segment j ($d(Sj)$), using Eq. (5.3).

$$\alpha = \min(d(Sj))\text{where } d(Sj) = \sqrt{\sum_{i=1}^{n}(x_i - y_i)^2} \tag{5.3}$$

x_i is the amplitude of i-th sample point of j-th segment obtained by experimentation, y_i is the amplitude of mutated chromosomes of i-th sample point of the j-th segment and n is the number of samples in the j-th segment, where $j >= 3$.

Offspring having higher fitness value are chosen for the next generation. The procedure continues for a large number of generations (say, 10 k generations).

5.2.6 Proposed Algorithm

The pseudo representation of the proposed algorithm is given stepwise.

```
Step 1  : n := total number of acoustic files
Step 2  : in_1 := fragment_1                    //first acoustic file
Step 3  : in_2 := fragment_2                    //second acoustic file
Step 4  : pos := position (fragment_2)      //find the position of the fragment
Step 5  : while ( pos <= n ) do
Step 6  :        cross_point := log2(x+pos)
Step 7  :        mut_point := log2(x+pos)
Step 8  :        offspring_1=CROSSOVER(in_1,in_2,cross_point)
Step 9  :        offspring_2=MUTATION(offspring_1,mut_point)
Step 10 :        if ( offspring_2 == in_3) then
Step 11 :          break
Step 12 :        else
```

Step 13 : cross_point : = MOVE1(cross_point) // get new cross point
mut_point := MOVE2(mut_point) // get new mutation point
Step 14 : end if
Step 15 : store (cross_point, mut_point) // store into the database
Step 16 : in_1 : = in_2
Step 17 : in_2 := in_3
Step 18 : repeat while
Step 19 : end of program

Following Example narrates the steps of the proposed GA based algorithm.

Example

CH 1: 10110010
CH 2: 11000001

After applying one-point crossover, say, at bit position 4, off-springs are generated using parent chromosomes CH1 and CH2:

Offspring1: 10110001
Offspring2: 11000010

Flip the value of the chosen gene, say at bit position 4 of the chromosomes, which produces the mutated off-springs as follows:

Offspring1: 10100001
Offspring2: 11010010

Select one of the off-spring which is best fitted. For the above example, the following outputs are obtained:

Offspring1 = 10100001 is scaled down as 33/127 = 0.2598.
It is a negative number and represented as: – 0.2598

$$\alpha = 0.57$$

Therefore, f (offspring1) = 1/exp(−0.57* – 0.2598) = 0.869.
Offspring2 = 11000010 is scaled down as 66/127 = 0.5196.
It is a negative number and represented as – 0.5196.
Therefore, f (offspring2) = 1/exp(−0.57* – 0.5196) = 0.744.
So, offspring12 is selected and forwarded to the next generation.

5.2.7 *Results and Discussion*

All the necessary information of the GA operators are stored within the database during execution of the algorithm. In the next step, the mill starts with a particular raw material (input mix) whose initial *psd* is known. A real-time recording is also started when the system starts crushing operation and the acoustic segments are extracted based on one full cycle of the vessel. The proposed model calculates at what state or after how many different full mill cycles the system needs to be stopped using the information (previous two consecutive segments, crossover point, mutation point, fitness value) stored in the database. Accordingly, a counter is set and whenever it reaches, after its desired count the system is stopped. The system also gives information about the target *psd* of the breakage material when the mill comes to an end, thus preventing wastage of materials.

The results of each step are given, as discussed above, for a particular type of raw material, i.e. Iron ore. A database for crushing of Iron ore is shown in Table 5.2 consisting of a priori information about the GA operators (crossover point and mutation point), states at which the mill needs to be stopped and corresponding two previous consecutive segments are given as input. Here, Segment 1 (S1) means the first full vessel cycle of the mill when the crushing is started, which is segmented from the captured acoustics of the mill. This way, Segment 2 (S2) is the next acoustic cycle and so on. After applying GA on two consecutive segments, the off-spring produced is considered as the simulated output. Similarity between the simulated output and the experimental output in terms of error also stored in the database during execution of the proposed GA based algorithm.

Figure 5.5a, b, c and d show the representation of segment no. 1 (S1) and segment no. 2 (S2), simulated output of segment no. 3 (S3) and experimental output of segment no. 3 (S3), respectively. Figure 5.6a, b and c show the plotting of the simulated and experimental result of a particular segment for three different running states, where the blue colour represent the experimental segment and the red coloured representation for the simulated segment. From the diagram it has been observed that there is a very negligible difference supplementing the performance of the proposed method. The L2–norm is calculated to find the error between the simulated and experimental results for a large number of data sets. The average error is of order 0.07.

5.3 Gaussian Mixture Model-Fuzzy C-Means Method

The aim of the Gaussian Mixture Model-Fuzzy C-Means (GMM–FCM) [15] based method is to predict the *psd* of the raw material with input mix at different grinding states of the running ball mill in a closed-loop operation [16]. As discussed earlier, first, the acoustics of the running ball mill in its different load-varying conditions are captured and segmented based on one full cycle of the mill vessel and on the

Table 5.2 Experimental inputs and simulated parameters

Experimental segment 1	Experimental segment 2	Simulated output for the experimental segment	Cross over point	Mutation point	Estimated error between experimental and simulated results
S1	S2	S3	4	5	0.077
S2	S3	S4	4	5	0.08
S3	S4	S5	4	6	0.067
S4	S5	S6	9	5	0.09
S5	S6	S7	8	4	0.065
S6	S7	S8	4	4	0.067
S7	S8	S9	5	5	0.07
S8	S9	S10	4	4	0.066
S9	S10	S11	4	4	0.056
S10	S11	S12	6	6	0.067
S11	S12	S13	4	5	0.071
S12	S13	S14	4	5	0.065
S13	S14	S15	4	5	0.046
S14	S15	S16	4	5	0.06
S15	S16	S17	4	4	0.066
S16	S17	S18	4	4	0.07
S17	S18	S19	4	4	0.065
S18	S19	S20	7	5	0.063
S19	S20	S21	5	4	0.063
S20	S21	S22	5	4	0.067
S21	S22	S23	8	9	0.072
S22	S23	S24	7	4	0.071
S23	S24	S25	2	7	0.076
S24	S25	S26	4	4	0.07
S25	S26	S27	7	5	0.06
S26	S27	S28	4	4	0.07
S27	S28	S29	4	4	0.066
S28	S29	S30	5	4	0.069
S29	S30	S31	4	5	0.066
S30	S31	S32	8	5	0.065

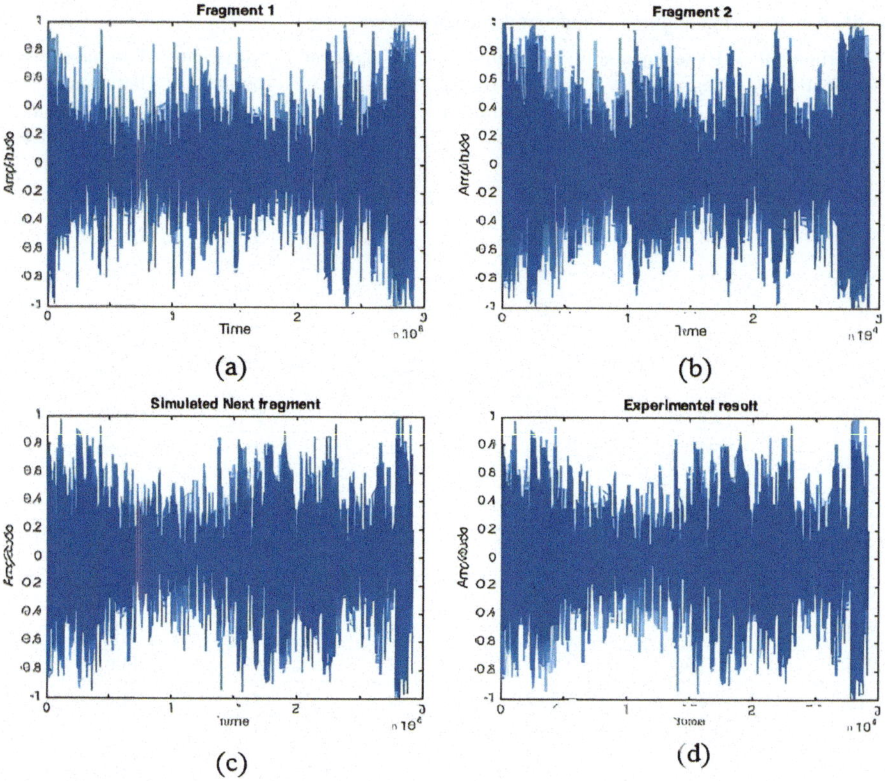

Fig. 5.5 Result using the proposed Genetic Algorithm

transition cycles. Finally, the database is created with the raw data, i.e. amplitude, generated from the segmented time domain acoustic signals.

For each different type of material, the ball mill runs from the beginning of crushing till the expected fines are produced. The crushed material is screened in a specific interval during crushing operation, and the corresponding *psd* of each state of grinding is stored separately. The aim here is to analyse the ball mill acoustics to predict the size distribution of the particles at any state of grinding. To carry out the experiment, the full cycles are considered while running the ball mill for different input mixes. GMM is applied to find the distribution of the data that is multivariate [17] in nature. Then using Fuzzy C-Means (FCM) clustering algorithm, clusters are generated where each cluster has different size ranges of the final crushed products. Figure 5.7 shows the block diagram of the proposed method. It has mainly two different steps:

(i) Find the data distribution of the acoustics using GMM.
(ii) Cluster the data using FCM.

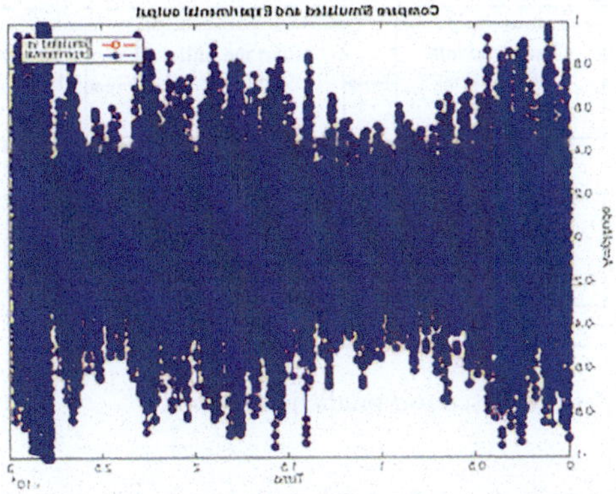

(a) : Comparison between the Simulated and Experimental Result of State1

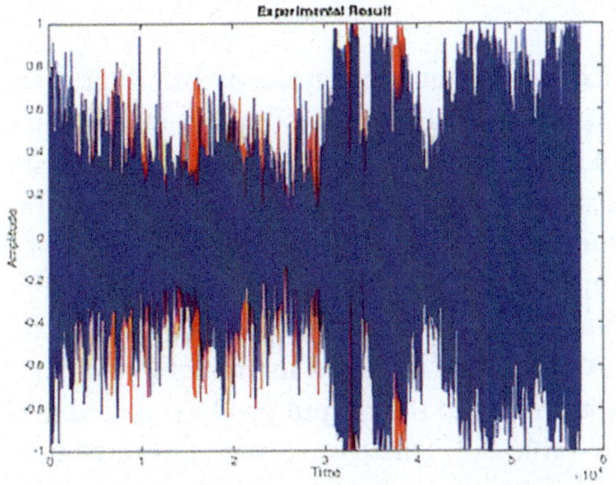

(b) : Comparison between the Simulated and Experimental Result of State2

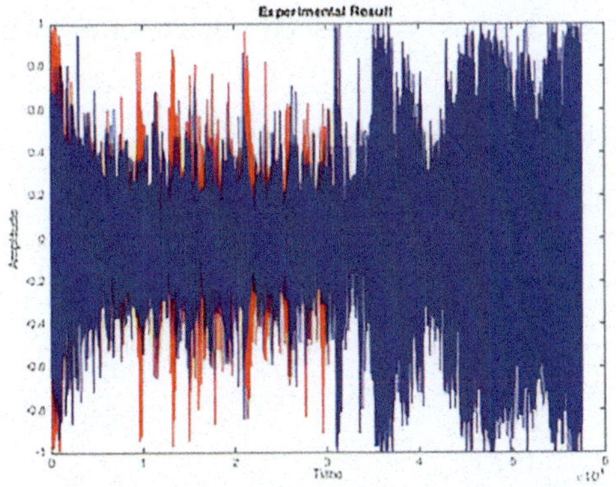

(c) : Comparison between the Simulated and Experimental Result of State3

Fig. 5.6 **a** Comparison between the simulated and experimental result of State1. **b** Comparison between the simulated and experimental result of State2. **c** Comparison between the simulated and experimental result of State3

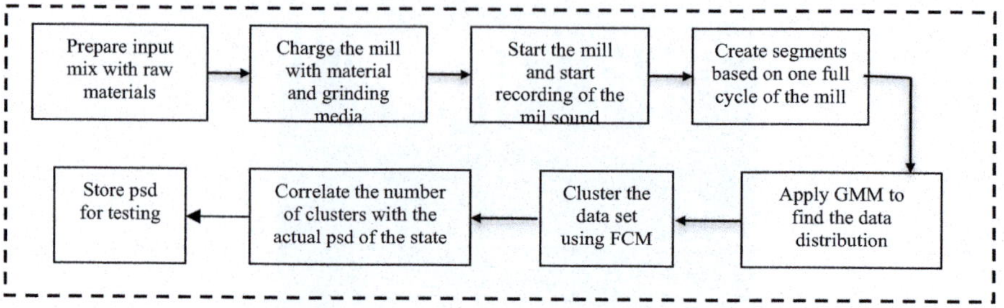

Fig. 5.7 Proposed GMM-FCM based *psd* prediction method

5.3.1 Gaussian Mixture Model (GMM)—Multivariate Data Distribution

It is mentioned in Chap. 4, that the nature of the data distribution is Gaussian. Depending on the expected final particle size, the input material size is predicted. The input provided in the vessel for grinding is considered as the mixture of different sized particles of the same material. It is not possible to find the size of each element present in the input mix and that is why the size distribution of the particle is considered for the measurement. During the grinding operation that size distribution is obviously changing and finally, the mill stops when the desired distribution of the particle is achieved. An example is shown in Table 5.3 for Iron Ore, as raw material. The feed size i.e., input mix and the output product size obtained after the grinding operations is given in Table 5.3. Due to crushing and screening operations there is a chance of material loss.

where the weight of the used grinding media is considered 2.7 kg and the crushing time 5 min. From Table 5.3, it has been observed that it follows a multivariate structure

Table 5.3 Feed and product size distribution

Particle size (mm)	Feed size weight (gm)	Product size weight (gm)
+ 5.6	313	436.9
+ 4.75	40	210.8
+ 1.70	37	233.4
+ 1.18	202.6	202.1
+ 850 (mm)	203.6	19.5
+ 710 (mm)	197.1	0
+ 500 (mm)	203.5	22.2
+ 300 (mm)	206.7	27.2
+ 90 (mm)	218.4	49.6
+ 75 (mm)	200	19.2
+ 45 (mm)	196.5	1

and GMM is applied on the segmented time series data to find the weighted sum of the data distribution.

The GMM is a probabilistic model used for representing a dataset as a mixture of multiple Gaussian distributions [18]. It assumes that the data points are generated from a combination of several Gaussian distributions with unknown parameters, such as mean and variance. GMM [19] is typically trained using the Expectation-Maximization (EM) algorithm, which iteratively estimates the parameter values of different distributions. First, the raw data of the segmented acoustic signals are taken as the input. Say, the 1-D representation of the input array is $X = \{x_1, x_2, \ldots..x_N\}$, where N is the number of data points. Next, the parameters of different Gaussian distributions are estimated to fit the patterns [20–22]. Three parameters need to be represented as

$$\theta = \{\alpha_k, \mu_k, \Sigma_k\},$$

where α_k, are the mixture weights, such as $k = 1 \ldots N$

μ_k is the mean vector
Σ_k is the covariance matrix.

Using expectation maximisation principle, a hidden variable is introduced in such a way that its knowledge would simplify the maximisation of likelihood estimation. At each iteration, following steps are performed:

- Estimate the distribution of the hidden variable given in the data and the current value of the parameters.
- Maximize the joint distribution of the data and the hidden variable.

The final distribution is obtained and stored in a 1-D array, say Y.

5.3.2 Data Analysis Using Fuzzy C-Means Clustering Algorithm

Fuzzy C-means (FCM) [23] clustering algorithm is applied on the data distribution, to find the number of clusters, generated at each state of grinding. Each data point may belong to multiple clusters, say 'C' number of clusters with varying degree of membership, rather than assigning each point to a single cluster, which is called hard clustering. It is based on the concept of fuzzy set theory, where each data point has been initialized with membership value between 0 and 1 for every cluster, indicating the degree of association. FCM minimizes an objective function, defined as the distance of each data point in terms of membership value from the cluster centroids and the membership values are updated. This algorithm is particularly useful in applications where data clusters are not well-separated, such as medical image segmentation, pattern recognition, and recommendation systems. Its iterative nature ensures the refinement of cluster centroids and membership values until convergence, making it a versatile tool for analyzing ambiguous or overlapping data sets.

If the number of clusters generated using simulation is the same as the number of particle size ranges generated during experiments, then it can be inferred that the predicted output follows the experimental results. In this step, the desired particle size distribution of the material is predicted. Gradually, with crushing, the distribution of the material is changing. So, in between the crushing procedures, to increase the knowledge base, the mill needs to be stopped and checks the produced material size distribution by screening the raw material and the corresponding information is stored in the database. Using FCM, the *psd* from the mill acoustics is found and tried to correlate with the actual *psd* stored during the experiment.

Let $Y = \{y_1, y_2, y_3 \ldots, y_n\}$ be the set of data points at any stage and $V = \{v_1, v_2, v_3, \ldots, v_c\}$ be the set of initial cluster centres. Using the Fuzzy partition cluster, centroids are calculated and updated until there is no change in the centroids. Information about the cluster is stored in a table for testing. So, the database is created based on the information gathered from GA and GMM-FCM operations.

5.3.3 *Simulated Outcome*

Gradually, when the crushing operation is carried away, the particles are grinded and the distributions are changed and generate different sizes of particles which differ from the input mix, given at the beginning of crushing. Figure 5.8 shows the representation of the clusters and the corresponding size distribution of the raw material at the beginning of crushing. In between the crushing operation, another state is also considered and the corresponding size distribution information is shown in Fig. 5.9. At the end of crushing the generated input mix size distribution is once again going to be changed, and the corresponding cluster and the distribution is shown in Fig. 5.10. Experimental output is compared with the simulated one and it was found that the simulated output is very much following the experimental results.

5.4 Summary

The grinding operation in a ball mill is a fully closed loop operation, from outside nothing can be assumed. To find the grinding product, the mill needs to stop its working and be screened to get information about the particle's size distribution. The procedure discussed in this section addresses that issue and can predict the size distribution of the particle without screening the material physically.

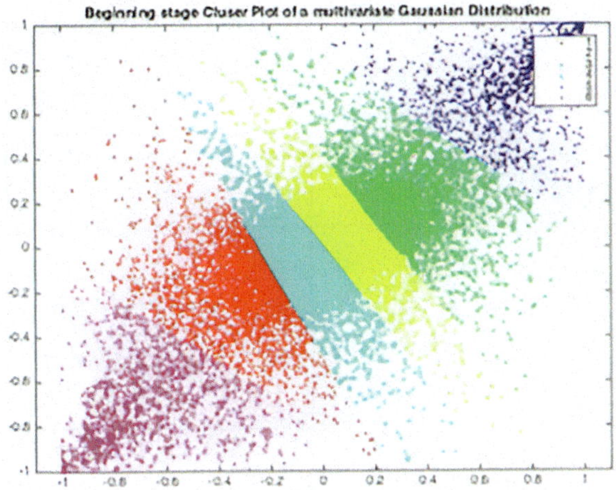

Mesh Size/ Diameter	Cluster No.
6 mm	1
14 mm	2
20 mm	3
40 µm	4
80 µm	5
150 µm	6

Fig. 5.8 Clusters and the *psd* at the beginning of crushing

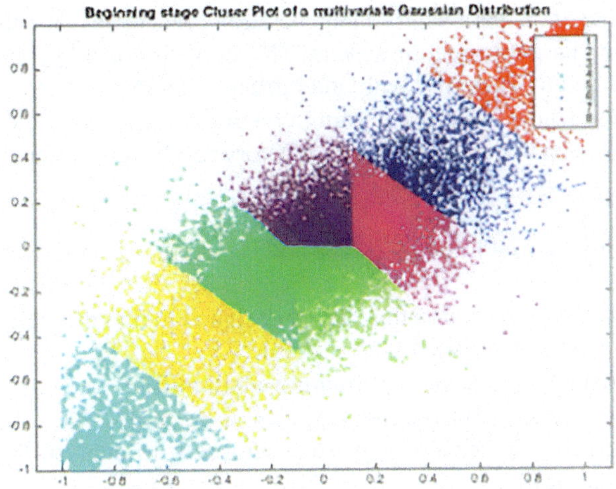

Mesh Size/ Diameter	Cluster No.
7 mm	1
20 mm	2
35 mm	3
55 mm	4
80µm	5
90 µm	6
120 µm	7
180 µm	8

Fig. 5.9 Different particle sizes and clusters after 2 min of crushing

The proposed GA-based method decides how many more cycles the mill needs to rotate from the current state to achieve the desired *psd* and to stop the mill. Using the GMM-FCM based method correlates the psd of different states of grinding.

In both the methods the predicted output follows the experimental result. So this prediction technique can be used in real time to control the mill operation. Acoustic evolution is an integral part of the work. Its importance basically lies with

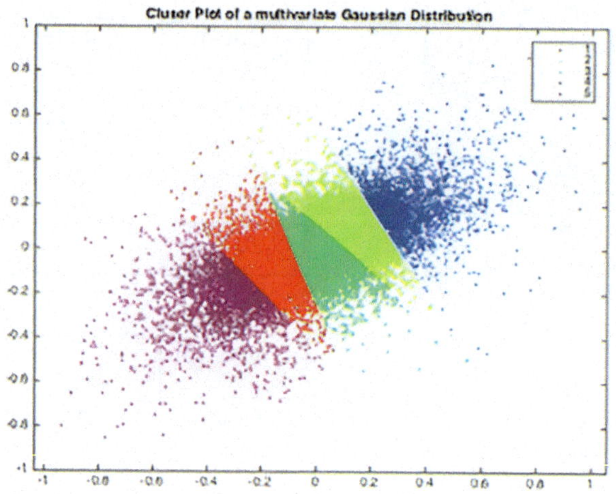

Cluster No.	Mesh Size/ Diameter
1	20 mm
2	80 mm
3	100 µm
4	200 µm
5	220 µm

Fig. 5.10 Different particle sizes and cluster plot at the end of crushing

the formation of the knowledge of the parameters and generation of feedback signals that is the key factor in a closed loop information system. The model is not only valid for the ball mill but also expected to work well with other different kinds of reactors/ crushers/compound machines. Under different training conditions convergence of error in acoustic signature has been investigated and satisfactory results are obtained.

Exercise

1. How the Genetic Algorithm is used to find the states of a running ball mill from acoustics?
2. Is there any better possible fitness function that can be designed to find the states of the running ball mill from its acoustics?
3. Why is the need of predicting the next state of running a ball mill?
4. How does the population is generated from the acoustics of the mill?
5. Why do we need to use Fuzzy C Means clustering algorithm to analyse the acoustics of the running ball mill?
6. Differentiate KDE and GMM.
7. How do we correlate the particle size distribution with the states of the running ball mill?

References

1. Andrade-Romero, J.A., Romero, J.F., Améstegui, M.: Acoustic signal characterization of a ball milling machine model. J. Phys. Conf. Ser. **285**, 012047 (2011). https://doi.org/10.1088/1742-6596/285/1/012047

2. Zhou, J.X.: Mill load control system based on chaotic particle swarm neural network. In: 2016 16th International Conference on Control, Automation and Systems (ICCAS 2016), pp. 778–781. HICO, Gyeongju, Korea
3. Herbst, J.A., Robertson, K., Rajamani, K.: Mill speed as a manipulated variable for ball mill grinding control. IFAC Autom. Min. Miner. Met. Process. **16**(15), 153–160 (1983)
4. Bhaumik, A., Sil, J., Maity, S., Das, T.: Designing an intelligent expert control system using acoustic signature for grinding mill operation. In: International Conference on Industrial Technology (ICIT 2006), IEEE Catalog Number 06TH8924C (2006). ISBN 1-4244-0726-5, Library of Congress, 2006935034
5. Cho, K.-S., Kim, S.-H., Lee, Y.: Correlation between acoustic intensity and ground particle size in alumina ball mill process. J. Korean Ceram. Soc. **55**(3), 275–284 (2018). https://doi.org/10.4191/kcers.2018.55.3.06
6. Bodenhofer, U.: Genetic algorithms: theory and applications. In: Lecture Notes, 3rd edn, pp. 11–31 (2004)
7. Man, K.F., Tang, W.K.S.: Genetic algorithms for control and signal processing. Ind. Electron. Control. Instrum. IECON 97 **4**, 1541–1555 (1997)
8. Bond, S., Hoeffler, A., Temple, J.: GMM estimation and empirical growth models. SSN 0265-8003, Discussion Paper No. 3048, pp. 1–30 (2001)
9. Dunn, J.C.: A fuzzy relative of the ISODATA process and its use in detecting compact well-separated clusters. J. Cybern. **3**(3), 32–57 (1973). https://doi.org/10.1080/01969727308546046
10. Bezdek, J.C., Ehrlich, R., Full, W.: FCM: the fuzzy c-means clustering algorithm. Comput. Geosci.. Geosci. **10**(2–3), 191–203 (1984). https://doi.org/10.1016/0098-3004(84)90020-7.ISSN0098-3004
11. Huang, C.W., Lee, K.C.: Application of ICA technique to PCA based radar target recognition. Prog. Electromagn. Res. **105**, 157–170 (2010)
12. Shi, Z., Liu, Y., Li, Q.: Medical image segmentation based on FCM and wavelets. In: Intelligence Science and Big Data Engineering. IScIDE 2013. Lecture Notes in Computer Science, vol. 8261. Springer, Berlin, Heidelberg (2013). https://doi.org/10.1007/978-3-642-42057-3_36
13. Umbarkar, A., Sheth, P.: Crossover operators in genetic algorithms: a review. ICTACT J. Soft Comput. **6**(1), 6 (2015). https://doi.org/10.21917/ijsc.2015.0150
14. Remde, S., Cowling, P., Dahal, K., Colledge, N.: Evolution of fitness functions to improve heuristic performance. In: Maniezzo, V., Battiti, R., Watson, J.P. (eds.) Learning and Intelligent Optimization. LION 2007. Lecture Notes in Computer Science, vol. 5313. Springer, Berlin, Heidelberg (2008). https://doi.org/10.1007/978-3-540-92695-5_16
15. Li, Y., Wang, Z., Zhao, T., Wanqing, S.: Research on a pattern recognition method of cyclic GMM-FCM based on joint time-domain features. IEEE Access **9**, 1904–1917 (2021). https://doi.org/10.1109/ACCESS.2020.3045815
16. Yu, L., Su, Z.: Application of kernel density estimation in lamb wave-based damage detection. Math. Probl. Eng.Probl. Eng. **2012**(1), 406521 (2012). https://doi.org/10.1155/2012/406521
17. Shiker, M.: Multivariate statistical analysis. Br. J. Sci. **6**, 55–66 (2012)
18. Saeidi, R., Sadegh Mohammadi, H.R., Amirhosseini, M.K.: An efficient GMM classification post-processing method for structural Gaussian mixture model based speaker verification. In: 2006 IEEE International Conference on Acoustics Speech and Signal Processing Proceedings (2006). https://doi.org/10.1109/ICASSP.2006.1660169
19. Hussain, H., Salleh, S.H., Ting, C.M., Ariff, A.K., Kamarulafizam, I., Suraya, R.A.: Speaker verification using Gaussian mixture model (GMM). In: Osman, N.A.A., Abas, W.A.B.W., Wahab, A.K.A., Ting, H.N. (eds.) 5th Kuala Lumpur International Conference on Biomedical Engineering 2011. IFMBE Proceedings, vol. 35. Springer, Berlin, Heidelberg (2011). https://doi.org/10.1007/978-3-642-21729-6_140
20. Reynolds, D.A., Quatieri, T.F., Dunn, R.B.: Speaker verification using adapted Gaussian mixture models. Digit. Signal Process. **10**(1–3), 19–41 (2000)
21. Muthusamy, H., Polat, K., Yaacob, S.: Improved emotion recognition using gaussian mixture model and extreme learning machine in speech and glottal signals. Math. Probl. Eng.Probl. Eng. **2015**(1), 394083 (2015)

22. Yessad, D., Amrouche, A.: SVM based GMM supervector speaker recognition using LP residual signal. In: Elmoataz, A., Mammass, D., Lezoray, O., Nouboud, F., Aboutajdine, D. (eds.) Image and Signal Processing. ICISP 2012. Lecture Notes in Computer Science, vol. 7340. Springer, Berlin, Heidelberg (2012). https://doi.org/10.1007/978-3-642-31254-0_66
23. Si, L., Wang, Z., Tan, C., Liu, X.: Vibration-based signal analysis for shearer cutting status recognition based on local mean decomposition and fuzzy C-means clustering. Appl. Sci. **7**(2), 164 (2017). https://doi.org/10.3390/app7020164

Chapter 6
Real-Time System Design to Predict the States of a Ball Mill

Abstract This Chapter discusses a research background to design a real-time Intelligent system to predict the states of crushing, including the End State of the loaded running ball mill. At the end of crushing, the mill stops its crushing operation, and the user obtains expected particle size distribution (*psd*). This Chapter proposes an intelligent system, which in real-time with some predefined control parameters achieves desired *psd* of a specific input mix, and stops the mill automatically. The system on chip can be designed so that the machine can be stopped automatically whenever the desired particle size distribution has been achieved. The proposed approach is based on the analysis of the time series data and validated using Machine learning technique.

6.1 Introduction

Grinding or crushing is an essential part of the mining industry to obtain the required particle sizes for different purposes. There are various other industries such as pharmaceutical, ceramic, paint etc. where sizes and weight of the fine products are very important for making the final product. Different particle size ranges (larger than the desired output) can be mixed for preparation of input mix by varying the speed of the dozer motor and also to limit the amount of dozing. The ball mill is charged with grinding media and with the input materials, the mill is allowed to start by switching on the motor attached to the mill. After a few minutes of crushing operation, the motor is stopped and the crushed particles are separated from the media. Manually, the materials are screened to classify different size ranges. User's specified particle size ranges are collected and rest of the materials, whose sizes are larger, fed again for further dozing and lower sizes of particles are stacked in the waste stack. This process will continue till to find two stacks of material; one is for desired output and another for wastage. In this manual procedure, a huge amount of undersized raw material is produced, which is nothing but a wastage.

To automate the entire process, existing methods of control engineering are studied but none found suitable to acquire desired sensed data due to operational hazard of

J. Sil et al., *Intelligent Real-Time Control Systems to Predict the Particle Size of Minerals Using Acoustics of Ball Mill*, Studies in Systems, Decision and Control 600,
https://doi.org/10.1007/978-3-031-91290-0_6

the mill, mainly dust. Artificial neural network (ANN) based classification, prediction and approximation models [1] are not new in the field of developing prediction systems. However, ANN operation is considered as black box and results are not interpretable. In the paper [2], the researchers proposed a dendritic neuron model which overcomes the drawbacks of ANN. The model is trained using different learning methods to obtain promising results in the field of classification, approximation and prediction. Deep learning techniques [3, 4], specifically convolution neural networks (CNN) [5] already proved to be the suitable techniques in designing autonomous systems. In this paper [6], the authors concentrated on three major architectural properties of CNN, i.e., the number of layers, filters, and filter size based on the performance. It has been noticed that the number of proper allocation of filters rather than an increase in the number of filters, largely affect the performance. Researchers provide a clear understanding in designing optimal network architectures using deep architecture. The paper [7] aims to model the energy consumption and effluent quality using multi objective optimal control in Wastewater treatment projects. Deep learning is applied for handling voluminous data and requires huge computational resources. CNN is suitable for image classification and selecting hyper parameters for pre-trained networks is very difficult to solve complex problems. Moreover, features are learnt from raw data, given as input and in most of the cases no preprocessing step is entertained. The noisy raw data, like acoustics of a ball mill, require preprocessing and features are extracted by analyzing the acoustics and perception of operators. Therefore, feature learning using deep architecture is costly due to inadequate large data sets.

In the last few years, researchers have been involved to analyze the operation of a ball mill based on surface vibration (commonly termed acoustic emissions) [8, 9] of mills. It is widely used as a non-invasive, low cost means of monitoring normally inaccessible attributes of processes. Acoustic emission [10, 11] method gives information about all the stages of the kinetic destruction process and expected effect of grinding. However, modelling using acoustic signature analysis for predicting the state of grinding is not reported yet. Few works are carried out to model and control the ball mill operation [12], though none of them are based on acoustic parameters. In the paper [13], grinding is optimized automatically based on the concept of mill vibration, not showing promising results. Due to the dust generated by the grinding media, prediction of particle size distribution using the sensor data is highly inaccurate. But, in this work the acoustics are considered, which have the prominent changes with changes of states of grinding and installed at the outside of the ball mill.

The aim of this Chapter is to propose an intelligent system which can predict states of running of the ball mill and correlate the weight of the particle size distribution in real time with the acoustics of the running ball mill. The experiment is conducted for a specific type of mill with a specific input mix with pre-defined control parameters, like grinding media type, size, mill speed etc. This system is designed based on the analysis of the time series data and validated using Machine learning technique. The system on chip can be designed in order to stop the machine automatically whenever the desired particle size distribution has been achieved.

Captured Acoustic signal is continuous in nature and the sound is changing with the crushing operation gradually with time, which is noticeable by human beings. Figure 6.1 shows one segment of each of the different states of the running ball mill in its load varying conditions where the raw material is considered as Coal. The time series signal's outer boundary, known as the envelope [14] is extracted from the acoustics signal. The envelopes are modeled using the auto regressive (AR) model [15] to predict the state of the running ball mill.

In another approach, the segmented waveforms are approximated using Gaussian distribution, applied on the raw data represented in time domain i.e. amplitude of the mill acoustics with respect to each segment. Figure 6.2 demonstrates that the distribution of amplitude, plotted for four different segments of the acoustic signal, are Gaussian in nature. The Kernel density function is employed to smooth out the contribution of each observed data point, considering the points belonging to its local neighborhood. Using both the methods, the prediction of the running states of the mill correlates with the particle size weight and the size distribution, obtained experimentally. A Feed Forward Neural Network [16] has been used in order to ensure that a trained model while developed using the experimental data performs well on unseen data and has good generalization capacity. The neural network model fits the nonlinear data and evaluates the performance of the system, satisfactorily.

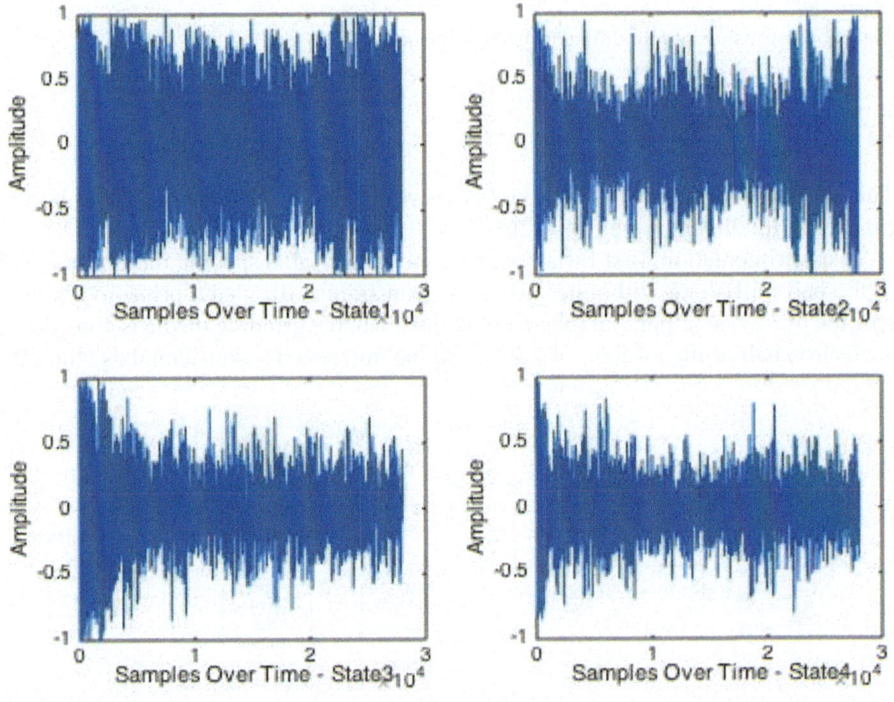

Fig. 6.1 Different states of running ball mill with raw material coal

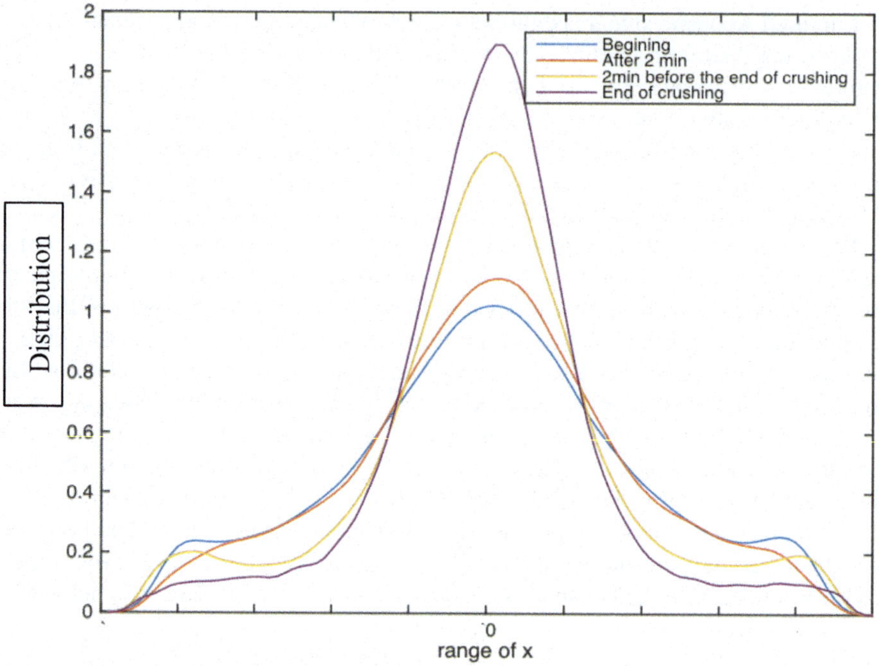

Fig. 6.2 Distribution of four different signal segments

6.2 Intelligent System

The workflow of the proposed model is shown in Fig. 6.3, which is framed in real time as an Intelligent system [24–26].

In experimentation, first for a particular raw material, a specific input mix is fed to the ball mill along with other necessary constraints, decided a priori to obtain a specific *psd* as an output. For the raw material Coal, the grinding media is considered as the iron balls with a weight of 2.7 kg and the mill speed is considered as 50 R.P.S.

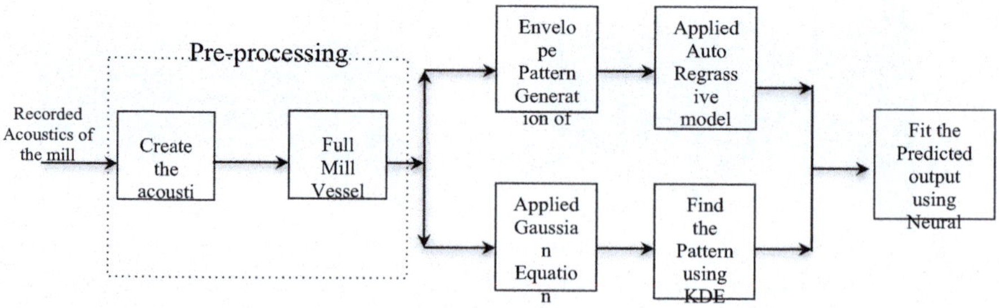

Fig. 6.3 Work flow of the proposed intelligent system

Two datasets are given in Tables 6.1 and 6.2 where the crushing time was considered as 5 min and 4.45 min, respectively. The input mix for both the cases with different particle size and weight is given in the 1st and 2nd column of the Tables 6.1 and 6.2, respectively. The third column of Tables 6.1 and 6.2 are dedicated to the weight of the grinded material.

Table 6.1 Feed and product size distribution of dataset 1

Particle size	Feed size weight (gm)	Product size weight (gm)
+ 5.6 (mm)	313	436.9
+ 4.75 (mm)	40	210.8
+ 1.70 (mm)	37	233.4
+ 1.18 (mm)	202.6	202.1
+ 850 (μm)	203.6	19.5
+ 710 (μm)	197.1	11.9
+ 500 (μm)	203.5	22.2
+ 300 (μm)	206.7	27.2
+ 90 (μm)	218.4	49.6
+ 75 (μm)	200	19.2
+ 45 (μm)	196.5	9.8

Weight of grinding media: 2.7 kg
Crushing time: 5 min
Mill speed: 50 R.P.S.

Table 6.2 Feed and product size distribution of dataset 2

Particle size	Feed size weight (gm)	Product size weight (gm)
+ 5.6 (mm)	436.9	420.3
+ 4.75 (mm)	210.8	211
+ 1.70 (mm)	233.4	218.9
+ 1.18 (mm)	202.1	198.8
+ 850 (μm)	19.5	200.2
+ 710 (μm)	11.9	10.1
+ 500 (μm)	22.2	18
+ 300 (μm)	27.2	41
+ 90 (μm)	49.6	85.8
+ 75 (μm)	19.2	17.6
+ 45 (μm)	9	9

Weight of ball: 2.7 kg
Crushing time: 4.45 min
R.P.S.: 50

After the first phase of grinding, the grinding media first removed and then the grinded raw material is brought out from the vessel to screen it using a sheave shaker to find the *psd* of the raw material and the weight of each sized particle. In the given example since the required *psd* is not achieved so, the same procedure is conducted again, termed as the second phase of grinding. The feed size (input) and product size (output) obtained from the second stage grinding process of the same raw material is given in Table 6.2.

The experiment is conducted based on the acoustics of the running ball mill in its different load varying conditions, and the acoustic is continuously captured. The acoustic is segmented based on the one full vessel cycle and the transition time cycle, then both are stored in the database for further use. This proposed method is based on one full mill cycle acoustic segment.

6.2.1 Finding Acoustic Signal Envelope

The signal envelope refers to the smooth curve that outlines the extremes of an oscillating signal over time. In signal processing, the envelope represents the varying amplitude of the signal, showing how it evolves. This concept is particularly important in analyzing and processing signals in audio, communications, and other areas of engineering and science. The envelope captures the maximum amplitude of the signal at each point in time and provides a measure of how the signal's strength changes over time.

It has been observed from Fig. 6.1 that the structure of the signal patterns for raw material Coal is different for different states of grinding. It has also been found during the experiment that, for different raw materials the structure of the acoustics at different states of grinding is different. So, with time when the grinding is performed within the mill, the acoustic curve outline is gradually changing which is clearly noticeable manually from outside. It is also observed that the amplitude range in reference to 0 is decreasing (in the upper side) or increasing (in the lower side) gradually with crushing and this is within a certain range. The amplitude range is in between [+ 0.8 to + 0.20] to [− 0.8 to − 0.20] for the raw material Coal, as shown in Fig. 6.1. So the aim is to extract the signal envelope of each state of grinding to find the uniqueness of each state directly from the captured acoustics. The signal envelopes are the curve tracing of the original signal y(t), which represents the boundary of the signal using an imaginary line. Assume a signal, defined using Eq. (6.1).

$$y(t) = a(t) * \cos[w_o * t + \text{Ø}(t)] \tag{6.1}$$

where, $a(t)$ and $\text{Ø}(t)$ contain only frequency components, while w_o represent the angular frequency.

The envelope $e(t)$ can be defined for the signal $y(t)$, given in Eq. (6.2).

$$e(t) = |a(t)| \tag{6.2}$$

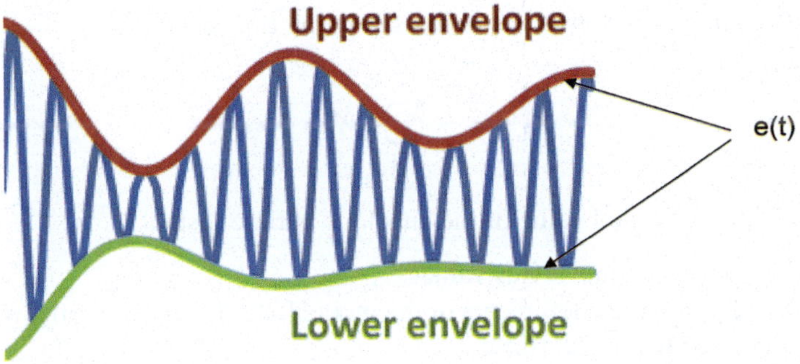

Fig. 6.4 Sample envelope

where $e(t)$ represents the envelope of the signal at time t, and $a(t)$ is a complex-valued function, denoted by its magnitude over time.

The envelope detector detects the upper and lower part of the signal boundary, from which the peaks of the signal are obtained. Figure 6.4 shows a small part of the sample envelope of a given signal, where the red line represents the upper envelope and the green line represents the lower envelope.

6.2.2 Autoregression Modelling

Auto regression (AR) is a statistical modeling technique used to describe time series data by regressing the current value of the series based on its previous values. In an autoregressive model, the current observation is expressed as a linear combination of past observations, plus a stochastic error term. The number of past observations included in the model is called the "order" of the autoregression. For example, an AR(1) model uses the immediately preceding value to predict the current value, while an AR(2) model uses the two preceding values. This assumes the past values and have a direct influence on the future values, making them particularly useful in analyzing time-dependent processes. In this procedure, the generated envelopes for each segment represent the randomness in the time series data. Autoregressive (AR) [15] model is employed in this work to fit the random pattern of each envelope using Eq. (6.3).

$$X_t = X_{t-1} + \varepsilon X_t = X_{t-1} + \varepsilon \tag{6.3}$$

Envelop at time t, X_t is represented using its previous time step, i.e. $t - 1$ and the smoothing factor is ε. An AR model predicts the future behavior based on the past when there exists correlation between the preceded and the succeeded values of the signal. It includes some degree of uncertainty or randomness, which models the acoustic signal very well.

The AR(p) model is defined as follows given in Eq. (6.4).

$$X_t = c + \sum_{i=1}^{p} \varphi_i X_{t-i} + \varepsilon_i \tag{6.4}$$

where, X_t is the predicted value of the envelope at time instant t,

X_{t-i} is the previous time series values,
$\varphi_1, \varphi_2, \ldots, \varphi_p$ are the coefficients of the model to fit the data, ε_i is the white noise and c is defined using Eq. (6.5).

$$c = \left(1 - \sum_{i=1}^{p} \emptyset_i\right)\mu c = \left(1 - \sum_{i=1}^{p} \emptyset_i\right)\mu \tag{6.5}$$

where μ is the mean of the signal.

The envelope is already extracted from the segmented signals for different raw materials. Now, using autoregression, the extracted envelopes of different states are modeled to define the patterns for the corresponding states, discussed in Sect. 6.2.1. The AR model is applied on the segments, obtained during grinding of a specific raw material from start till end. So, two sets of patterns of envelopes are obtained for the upper and the lower boundary of each segment after modeling all the captured segments. This method is applied to all the different samples of a particular input mix to regenerate the pattern of its running states. This helps to build a database, which can be used to correlate the particle size distribution with different states of the running ball mill dynamically. At different running conditions, the information about the *psd* and the weight of the material, shown in Tables 6.1 and 6.2, correlate the same for later use.

6.2.3 *Kernel Density Estimation*

Kernel Density Estimation (KDE) [17] is a non-parametric method used for estimating the probability density function (PDF) of continuous random variables. It has been already noticed that the acoustics of the running ball mill follows the Gaussian Distribution, discussed in Chap. 4. In this work, PDF of the raw data is calculated for each fragment using Eq. (6.6) and finds the corresponding peaks. Kernel density is applied on a dataset to obtain the pattern, which shows a specific peak. The pattern is used to classify the states of the running ball mill with different size distribution using a multilayer perceptron model [18].

Using that concept, here the aim is to do the prediction of the different states of the running ball mill at runtime within a minimum time frame. Here, in this proposed method a pattern is extracted using a probability density-based approach that is based on Gaussian distribution function. The acoustics of the ball mill with different raw materials in its load varying conditions, follows the Gaussian function, which is described by Eq. (6.6).

$$y = ae^{\left(\frac{-(x-b)}{c}\right)^2} \tag{6.6}$$

where a is the peak value, b is the position at which the peak occurs and c is the standard deviation, which controls the width of the curve, x is the time series data and y is the corresponding value obtained after applying Eq. (6.6) on the time series data.

6.2.4 Neural Network Model

A neural network is a computational model inspired by the structure and function of the human brain, designed to recognize patterns and solve complex problems. It consists of layers of interconnected nodes, or "neurons," where each node processes input data and applies a mathematical function to generate output. The network typically includes an input layer, one or more hidden layers, and an output layer. Neural networks are trained using large datasets and a process called backpropagation, which adjusts the weights of the connections to minimize errors in predictions. They are widely used in various applications such as image and speech recognition, natural language processing, and autonomous systems, due to their ability to model complex, non-linear relationships in data.

In this work, using AR, the envelope of a state of the running ball mill which is unique in nature and changing with time, is modeled and the corresponding parameters are set for different states. Also, the patterns obtained for different states using the Kernel density estimation are now available to classify the patterns for different states of the running ball mill. Neural network model is used to validate the prediction of particle size distribution with the respective class labels. Neural Network [22, 23] is used for function approximation and recognizing the patterns. Here, a two-layered feed-forward network [19], shown in Fig. 6.5 has been designed with sigmoidal activation [20] function at hidden layers, which maps the net input values to particular states, represented as output. Ten neurons in each hidden layer is fixed experimentally. The network is trained with the Levenberg-Marquardt back propagation algorithm [21], which can train the system very fast, but for a big data set it requires a huge memory. Here, the raw material Coal is considered for experimentation where the crushing continues till the desired *psd* is not achieved. The experiment is conducted with different raw materials and results are verified with the neural network model.

Consider an example, where the total number of segments generated is approximately 1800. So a sample set of 1800×12 is prepared where the sample size is 1800 and the number of the parameters for each sample is 12. The input feature vector of dimension 12 is applied at the neurons of the input layer, and 1800 such feature vectors are used to train the neural network model. The parameters that are

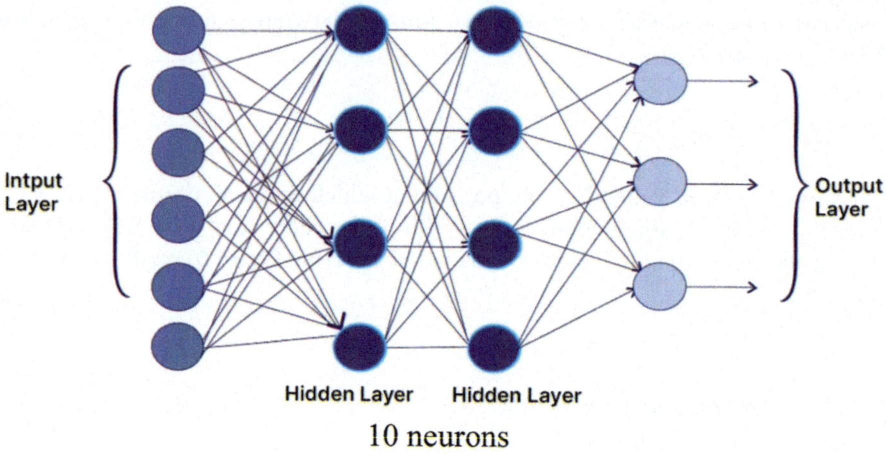

Fig. 6.5 Feed forward neural network classifying four grinding states

considered here are the different statistical moments—like mean, variance, standard deviation, skewness, kurtosis and all the second moments, which are already considered in one of the procedures, discussed in Sect. 4.3 of Chap. 4. Other than these statistical parameters, 'a', 'c', 'ε' are the coefficients of the AR model, forming the feature vector of dimension 12. The number of output neurons defining four class labels as the four equidistant states of the running ball mill, for a particular particle size distribution.

6.3 Simulated Outcome

This Chapter is dedicated to design a method which can be used in runtime to predict the state of grinding and take the decision to stop the ball mill. In this method different patterns are generated from the time domain data and based on that the prediction of the ball mill states can be taken. The Full_Cycle acoustics are considered as input at the beginning. As discussed in Sect. 6.1, at first the signal envelope is extracted from the acoustic segments of the running ball mill in its load varying condition. Envelope is the signature of the acoustic segments which signify the crushing that takes place within the mill.

The envelope is the boundary of each segment given as the input, represented as two vectors containing the maximum amplitude at each instant of time. The outline of the signal having upper and lower envelope of the segmented samples are shown in Fig. 6.4. From each of the envelopes of different segments of a particular raw material, the peak value for each boundary is extracted, and stored in two different arrays named Envelope_Lower_Peak and Envelope_Upper_Peak. If both the arrays are plotted in a single graph then it can be easily distinguishable from the pattern which is changing with time. Figure 6.6 shows the plotting of the Envelope_Lower_Peak and Envelope_Upper_Peak of around 98 samples which are extracted from the lower and upper envelope of the raw material Coal, to produce a specific *psd*. It is noticed that the lower and the upper envelope is gradually increasing and decreasing, respectively with the crushing of the material. It is also experimentally found that this pattern structure of the envelope is similar for all the different input raw materials, used for crushing.

Auto Regression is used to model the Upper and Lower envelope of each state. Figure 6.7 shows a fragment of upper envelope, simulated using AR consisting of both the experimental (in blue colour) and the simulated results (in red colour). From Fig. 6.7, it is noticed that the simulated output follows the experimental results. Using AR, the information of 1st stage of grinding, given in Table 6.1 is modelled, where the smoothing constant is $\varepsilon = 0.14$. A fragment of the regenerated segment is shown in Fig. 6.8. For each different states (discussed in Sect. 6.1) of the running ball mill with the raw material Coal, the regeneration of the envelope is done and the corresponding ε value is found, which is given in Table 6.3. It is clear from the value of ε that the parameter is gradually decreasing with time or towards the grinding.

After that KDE is applied to the acoustic segments of the full mill cycles to predict the unique pattern of the different states of grinding of a given raw material using

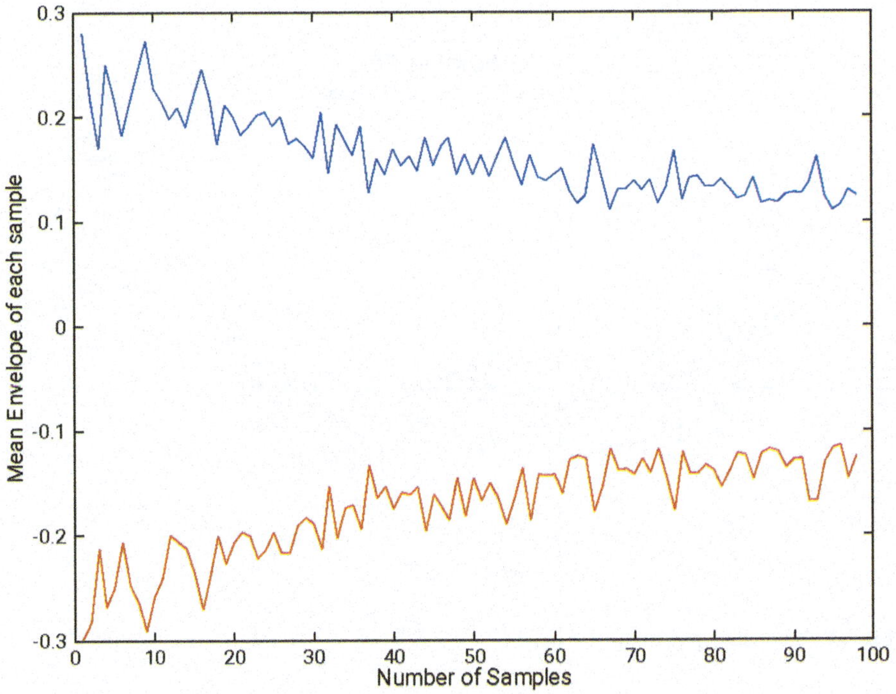

Fig. 6.6 Upper and lower envelope of the segmented samples

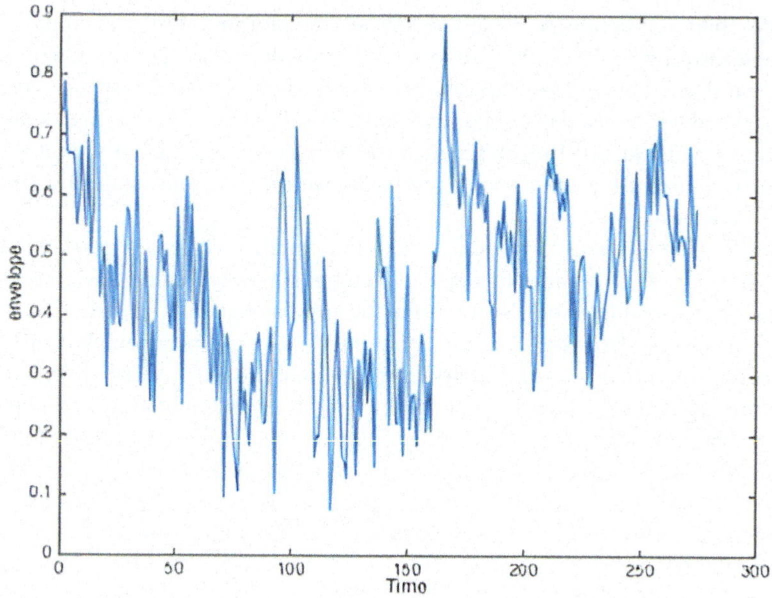

Fig. 6.7 Upper envelope simulated using autoregression

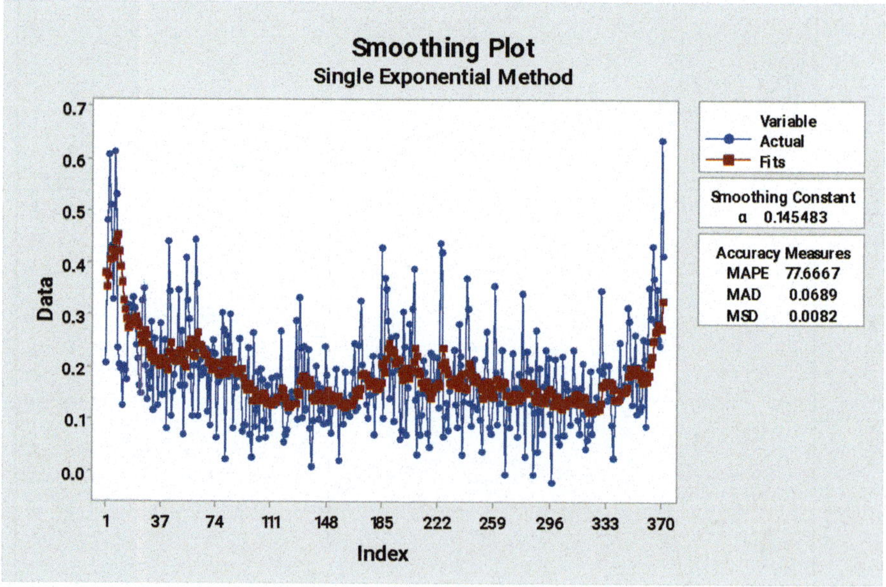

Fig. 6.8 Smoothing plot of the first stage

Table 6.3 Parameters used for prediction of the states

Parameters	Raw material	State1	State2	State3	State4
Smoothing parameter ε for AR	Coal	0.14	0.13	0.06	0.05
a for KDE	Coal	0.178	0.179	0.1	0.087
c for KDE	Coal	0.5376	0.4666	0.2904	0.266

Gaussian Equation and kernel density estimate. Gaussian distribution function, given in Eq. (6.6), is applied on each of the segmented data sets of a particular raw material and has found the kernel density estimate for the same.

The pattern of each of the running states and peaks are estimated and it is observed that the density pattern for each state is varying based on crushing. Figures 6.9, 6.10, 6.11 and 6.12 show the patterns generated using KDE and Gaussian Equations for four different states, as State1, State2, State3 and State4 of the ball mill, respectively during grinding and the density parameters are extracted.

It is noticed from Figs. 6.9 and 6.10 that the pattern and the peak value of the states, State1 and State2 are mostly the same. However, gradually with crushing the patterns are changing. In state State3 the pattern peak is totally different from the first two states. At the end of crushing at State4, the pattern is different from all the other states. It also experimented that this is true for all the different types of raw materials. Based on this signature of the acoustics it can be easily defined as the end of crushing of a particular raw material.

The proposed system is modeled for prediction of the states using AR and KDE, and the estimated parameters ε, a and c are given in Table 6.3. The parameters are

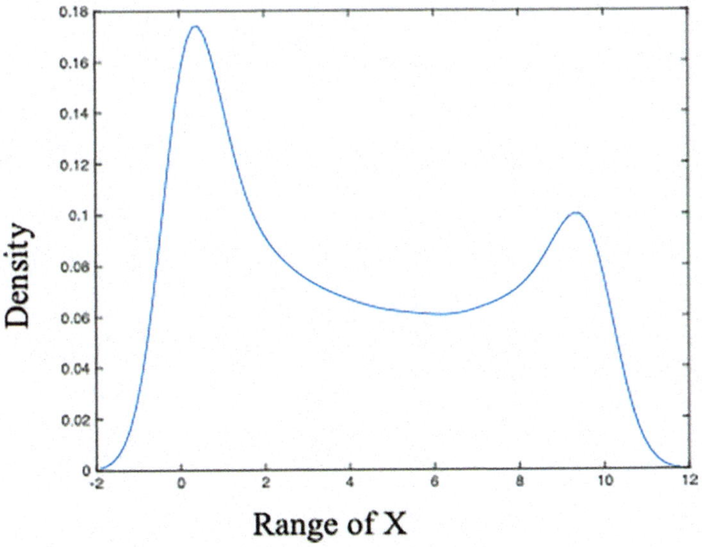

Fig. 6.9 Kernel density at the beginning of crushing (State1)

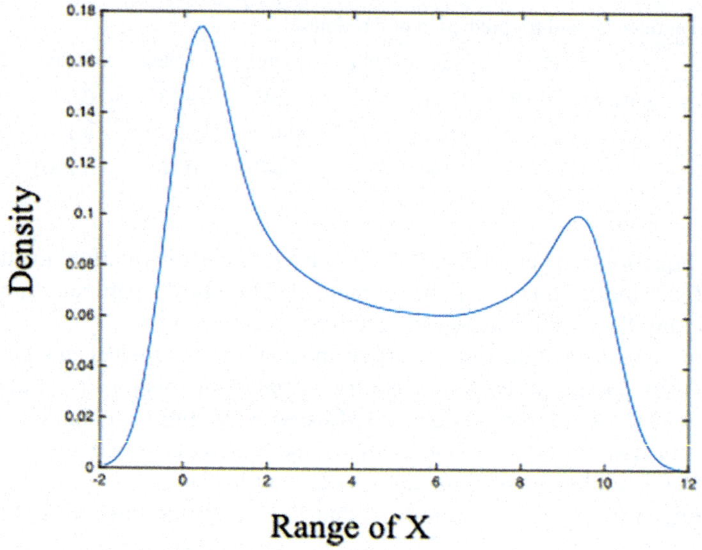

Fig. 6.10 Kernel density of the second state of the crushing (State2)

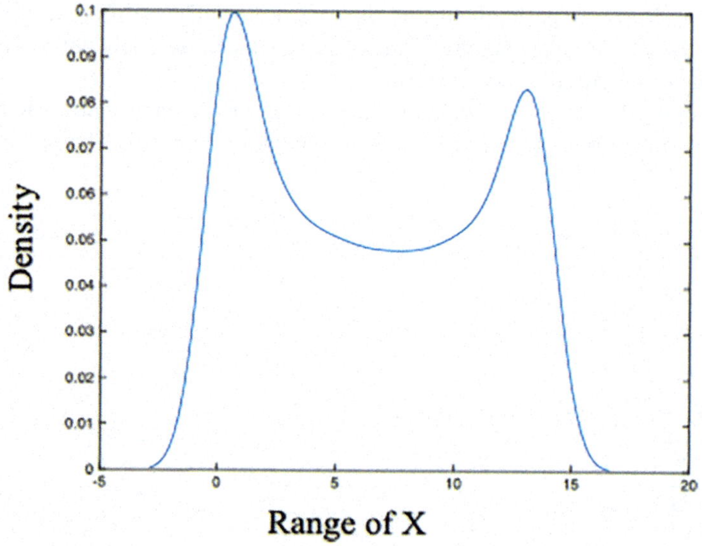

Fig. 6.11 Kernel density of the third state (State3)

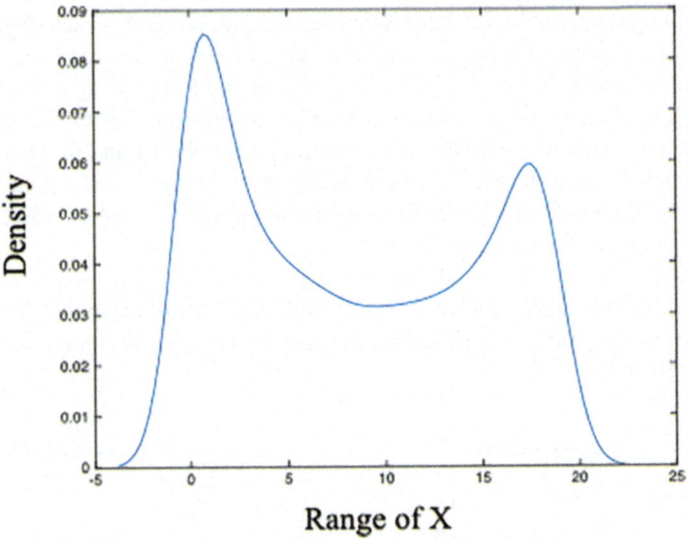

Fig. 6.12 Kernel density at the end of crushing (State4)

in descending order with crushing. All the related information is stored within the database for later use.

In the Auto Regression (with one parameter—AR(1)) process, with only the previous term in the process and the noise term (positive ε) contribute to generating the output. The noise term approaches one, the output depends on the previous term, resulting in "smoothing" of the output, which is similar to the effect of low pass filtering. From Table 6.3, it is clear that gradually with crushing the noise becomes the white noise. It has been observed that the value of 'b' is zero for all cases because it represents the centre of the curve and in KDE, the peak is always at the 0th value. The parameter c is the standard deviation. It has been observed from Table 6.3 that the value of c is gradually decreasing, corroborating the fact of functioning a ball mill. More specifically, c is higher at the beginning of crushing and gradually reaches to the minimum value when the mill completes the crushing operation. The peak of the curve is also gradually decreasing with the crushing as friction in between the materials is diminishing.

Finally, for validation, the neural network is trained using Coal as the material for experimentation and the captured acoustic is fragmented into 1800 different segments while the number of the extracted parameters are 12. So 1800 samples, each having feature vectors of dimension 12 are prepared and fed as input patterns to the neural network. The randomly chosen 60% input is used to train the network. Rest of the sample is used for testing to predict the states of the running ball mill. Here, the class labels are found as output which follows the states of the running ball mill. Figure 6.13 shows the results of the neural network. It has four different graphical representations, where Fig. 6.13a is the output that states how well the model is learnt on the training data. Figure 6.13b presents the validation output which mentions how well the model generalizes to unseen data, which helps to prevent overfitting. The test output is shown in Fig. 6.13c. Finally, a combined view is given in Fig. 6.13d for the dataset. The R-value of the model i.e., correlation coefficient for the training data set is approximately 0.99901, for validation it is 0.99761 and for testing dataset it is 0.99889. In each graph, the R-values are close to one and the data points are fitted around the lines, which indicate a strong linear relationship exists between the predicted output and the target output.

The error analysis for overall data points is shown in Fig. 6.14, where the mean square error is very small which confirms the effectiveness of the model. This demonstrates that the Neural Network Model has learnt to predict the target variable with very high accuracy.

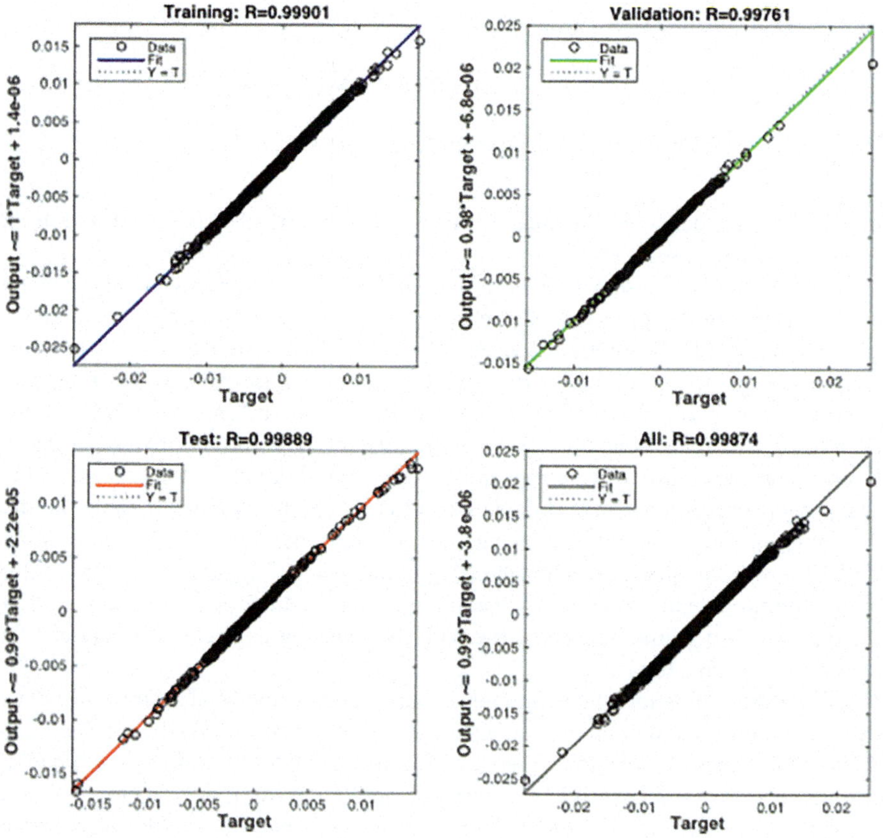

Fig. 6.13 Results of training and validation dataset using neural network

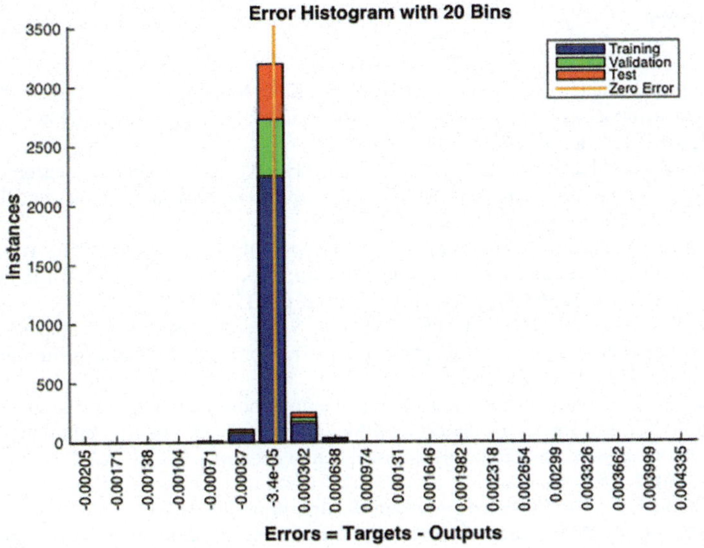

Fig. 6.14 Error analysis using neural network model

6.4 Summary

The aim of this work is to provide an automatic grinding system to predict the states of grinding and can correlate the same with *psd* and weight of the material. Though the ball mill has been taken for experimentation, the system can be ported to any other kinds of crusher just by formation of a training data set. Till there is no such work for predictive grinding and hence there is no autonomous system. Here each stage of the running ball mill (with different particle size distribution) can be defined with a simple Equation. Also another prediction strategy is proposed using Gaussian Equation and Kernel density estimate, which can provide the pattern for different running states. Using the Neural Network, the experimental results are validated. The simulated output is similarly behaved as the experimental result.

Exercise

1. What is a Signal Envelope? How to extract the envelope of an Acoustic signal?
2. What is the range of Amplitude of a running ball mill in its load varying condition?
3. Which method can be used to reconstruct the envelope of a signal?
4. Why is the Levenberg-Marquardt back propagation algorithm used to train the Network designed for state prediction of a running ball mill?
5. How many and what are the different parameters used to design a network for the running ball mill state prediction?
6. What are the applications of AutoRegressive models in time series analysis?
7. How does envelope analysis help in vibration and fault detection in mechanical systems?

References

1. Abiodun, O.I., Jantan, A., Omolara, A.E., Dada, K.V., Mohamed, N.A., Arshad, H.: State-of-the-art in artificial neural network applications: a survey. Heliyon **4**(11), e00938 (2018). https://doi.org/10.1016/j.heliyon.2018.e00938
2. Gao, S., Zhou, M., Wang, Y., Cheng, J., Yachi, H., Wang, J.: Dendritic neuron model with effective learning algorithms for classification, approximation, and prediction. IEEE Trans. Neural Netw. Learn. Syst. **30**(2), 601–614 (2018)
3. Gu, S.X., Holly, E., Lillicrap, T., Levine, S.: Deep reinforcement learning for robotic manipulation with asynchronous off-policy updates. In: Proceedings of 2017 IEEE International Conference Robotics and Automation, Singapore, pp. 3389–3396 (2017)
4. Kuutti, S., Bowden, R., Jin, Y., Barber, P., Fallah, S.: A survey of deep learning applications to autonomous vehicle control. IEEE Trans. Intell. Transp. Syst. (2019). https://doi.org/10.1109/TITS.2019.2962338
5. Welling, E.: Convolutional neural networks in autonomous vehicle control systems (2017)
6. Kebria, P.M., Khosravi, A., Salaken, S.M., Nahavandi, S.: Deep imitation learning for autonomous vehicles based on convolutional neural networks. IEEE/CAA J. Autom. Sin. **7**(1), 82–95 (2019)
7. Ranganathan, A.: The Levenberg-Marquardt Algorithm (2004)
8. Spencer, S.J., Campbell, J.J., Weller, K.R., Liu, Y.: Acoustic emissions monitoring of SAG mill performance. IEEE Explore (1999). https://doi.org/10.1109/IPMM.1999.791509
9. Pryadko, N.: Optimization of Fine Grinding on the Acoustic Monitoring Basis, Energy Efficiency Improvement of Geotechnical Systems. Pivnyak, Beshta & Alekseyev, Taylor & Francis Group, London, pp. 101–108 (2014). ISBN 978-1-138-00126-8
10. Desobry, F., Davy, M., Fitzgerald, W.J.: Density kernels on unordered sets for kernel-based signal processing. In: 2007 IEEE International Conference on Acoustics, Speech and Signal Processing-ICASSP'07, pp 417–421. IEEE. https://doi.org/10.1109/ICASSP.2007.366261
11. Yu, L., Su, Z.: Application of kernel density estimation in lamb wave-based damage detection. Math. Probl. Eng. **2012**(1), 406521 (2012). https://doi.org/10.1155/2012/406521
12. Lu, S., Zhou, P., Chai, T.: Modeling and simulation of whole ball mill grinding plant for integrated control. In: IEEE Transactions on Automation Science and Engineering (2014). https://doi.org/10.1109/TASE.2013.2296309
13. Wang, X., Lu, X., Jia, Z., Wang, W., Wang, F.: Control system for micro milling machine. Int. J. Ind. Syst. Eng. (2015). https://doi.org/10.1504/IJISE.2015.070876
14. Yang, Y.: A signal theoretic approach for envelope analysis of real-valued signals. IEEE Access (2017). https://doi.org/10.1109/ACCESS.2017.2688467
15. Dai, X., Liu, J., Zhang, H.: Application of AR model in the analysis of preearthquake ionospheric anomalies. Math. Probl. Eng. **2015**(157184), 5 (2015). https://doi.org/10.1155/2015/157184
16. McLean, D., Bandar, Z., O'Shea, J.D.: The evolution of a feedforward neural network trained under backpropagation. In: Artificial Neural Nets and Genetic Algorithms. Springer, Vienna (1998). https://doi.org/10.1007/978-3-7091-6492-1_114
17. Zucchini, W.: Applied Smoothing Techniques Part 1: Kernel Density Estimation, pp. 1–20 (2003)
18. Rosenblatt, F.: The perceptron: a probabilistic model for information storage and organization in the brain. Psychol. Rev. **65**(6), 386–408 (1958)
19. Sazli, M.: A brief review of feed-forward neural networks. Commun. Fac. Sci. Univ. Ank. **50**, 11–17 (2006). https://doi.org/10.1501/commua1-2_0000000026
20. Narayan, S.: The generalized sigmoid activation function: competitive supervised learning. Inf. Sci. **99**(1–2), 69–82 (1997). https://doi.org/10.1016/S0020-0255(96)00200-9.ISSN0020-0255
21. Reynaldi, A., Lukas, S., Margaretha, H.: Backpropagation and Levenberg-Marquardt algorithm for training finite element neural network. In: 2012 Sixth UKSim/AMSS European Symposium on Computer Modeling and Simulation, pp. 89–94 (2012). https://doi.org/10.1109/EMS.2012.56

22. Qiao, J., Zhou, H.: Modeling of energy consumption and effluent quality using density peaks-based adaptive fuzzy neural network. IEEE/CAA J. Autom. Sin. **5**(5), 968–976 (2018)
23. Wu, S.Q., Er, M.J.: Dynamic fuzzy neural networks—a novel approach to function approximation. IEEE Trans. Syst. Man Cybern. B (Cybern.) **30**(2), 358–364 (2000)
24. Mohammadi, H., Ghazanfari, M., Pishvaee, M.S., Teimoury, E.: Fresh-product supply chain coordination and waste reduction using a revenue-and-preservation-technology-investment-sharing contract: a real-life case study. J. Clean. Prod. **213**, 262–282 (2019). https://doi.org/10.1016/j.jclepro.2018.12.120
25. Sen, S., Bhaumik, A.K., Sil, J.: Prediction of the state of grinding materials in run time using genetic operator. In: Advances in Intelligent Systems and Computing Applications and Future Directions—Volume I, Advances in Intelligent Systems and Computing, vol. 798, pp. 475–486 (2017). https://doi.org/10.1007/978-981-13-1132-1_37
26. Shamsollahi, M., Badiee, A., Ghazanfari, M.: Using combined descriptive and predictive methods of data mining for coronary artery disease prediction: a case study approach. J. AI Data Min. **7**(1), 47–58 (2019). https://doi.org/10.22044/jadm.2017.4992.1599

Chapter 7
System Validation

Abstract In this Chapter, Petri-like net structure is used to develop an intelligent expert control system in order to achieve optimum wastage of the grinding material by regulating the parameters of the grinding mill. The work establishes an appropriate theoretical background that helps to predict dynamic breakage characteristics of the particle size distribution of materials, adequately supported by experimental data. A feed-forward neural network has been employed in this work to adapt the dynamic breakage characteristics of the mill by imparting training using the backpropagation learning algorithm. The mill has been tuned with the trained parameters and no further change in parameters is required so long the desired output remains the same, irrespective of the input particle size range. For reasoning the breakage characteristics of the mill, a rule base is designed using the recorded information of the acoustic signal and implemented using the Petri-like net structure. In real-time, the reasoning process generates inference in case there is any deviation from the recorded acoustic signal, which is subsequently converted to control action and achieving optimum wastage of grinding material. The time of reasoning is compatible with the particle residence time.

7.1 Introduction

The intelligent system has been developed for the operation of the ball mill, discussed in Chap. 6. However, before implementing such a system, modelling, simulation and stability analysis are very much required to claim about the full functioning of the ball mill. In this Chapter, the reasoning and learning methods are used for preparation of the knowledge base in terms of If-Then rules. Acoustic signatures have been used for building the If-Then rules whereas Petri–net structure is invoked for knowledge representation using the rules. The Petri–net structure [1, 2] is used for reasoning and analysing stability of the system.

A Petri-net is a directed bipartite graph consisting of places and transitions, used as an automated tool for reasoning with a rule based system. Uncertainty of the rules

J. Sil et al., *Intelligent Real-Time Control Systems to Predict the Particle Size of Minerals Using Acoustics of Ball Mill*, Studies in Systems, Decision and Control 600,
https://doi.org/10.1007/978-3-031-91290-0_7

are modeled by using Fuzzy-Petri_nets. In this work, the Extended Petri-net structure has been used for modelling and simulation of the intelligent system.

7.2 Inference Generation Model

The synergism between reasoning and learning techniques are exploited in the work to develop an intelligent expert control system using Petri-net structure. The model delineates the grinding operation of the ball mill and tunes the parameters according to the targeted particle size distribution. Two kinds of information: (i) particle size distribution and (ii) acoustic signatures of the mills under varied load conditions are required to determine the operational status of the ball mill. The entire scheme of the ball mill has been described using the rule base of the tumbling mill and realized by designing an extended-Petri-net (EPN) structure. The operational status of the ball mill has been investigated using the reasoning process. The inferences generated from the reasoning process are converted into control actions in order to achieve the desired product.

7.2.1 Extended Petri-Net Model

Definition 1. An EPN is a directed bipartite graph [3] with ten tuples denoted by

$EPN = \{P, D, n, T, CT, A, tt, ctt, I, O\}$.
$P = \{p_1, p_2, \ldots\ldots, p_z\}$ is a set of places, represented by "circles", where $1 \le i \le z$,
$D = \{d_1, d_2, \ldots\ldots, d_z\}$ is a set of variables, each d_i is mapped to the corresponding p_i, where $1 \le i \le z$,
n: $P \rightarrow [0\ 1]$, is an association function, represented by a mapping from a place to a binary number. n_i denotes the truth value of variable d_i, mapped at place p_i.
$T = \{t_1, t_2, \ldots\ldots, t_m\}$ is a set of transitions, represent the '*if–then*' part of the rules and denoted by "bars", where $1 \le i \le m$,
$CT = \{ct_1, ct_2, \ldots\ldots, ct_m\}$; is a set of negate transitions. A variable and its complement are mapped at the input and output places of a negate transition, respectively and denoted by "parallel bars", where $1 \le i \le m$.
$A \subseteq (PxT) \cup (TxP) \cup (PxCT) \cup (CTxP)$ is the flow relations and the elements of A are called 'arcs'.
tt: $T \rightarrow [0\ 1]$, is an association function, represented by a mapping from transition to a binary number where tt_i denotes the firing condition of the rule mapped at transition t_i.
ctt: $CT \rightarrow [0\ 1]$, is an association function, represented by a mapping from negate transition to a binary number where ctt_i denotes the firing condition of transition ct_i.

I and O are two association functions which represent the mapping from transitions or negate transitions to their input and output places, such that PxT $\subseteq$ A, TxP $\subseteq$ A or PxCT $\subseteq$ A, CTxP $\subseteq$ A, respectively.

A negate transition [4] has been introduced in the proposed Petri-like-net-structure to represent a variable and its complement as input and output places of the negate transition, respectively. The negate transition fires if and only if its input place holds truth value ('1' or '0'), propagated from its predecessor places. After firing, truth value at the output place of a negate transition always is the complement of the truth value of its input place.

The parameters introduced above may be illustrated with the following example.

Example 1. Let us consider the following If-Then rules.

R1: If (the_motor_has been_properly_initialized) Then (the_motor_is_ok).

R2: If (the_motor_is_not_ok) Then (reinitialized_the_motor).

R3: If ((the_motor_is_ok) And (dozer_is_ok)) Then (dozer_is_loaded).

The EPN in Fig. 7.1 describes the above rules [5, 6].

Here, in rule *R1*, variable d_1 (the_motor_has_been_properly_initialized) and d_2 (the_motor_is_ok) are mapped at place p_1 and p_2 respectively. Variable d_3 (the_motor_is_not_ok) and d_4 (reinitialized) of *R2* are mapped at places p_3 and p_4 respectively. Similarly, variable d_5 (dozer_is_ok) and d_6 (dozer_is_loaded) of rule *R3* are mapped at place p_5 and p_6 respectively. Variables d_2 and d_3 are complemented with respect to each other and mapped as input and output places of the negate transition ct_1.

The list of parameters used in Fig. 7.1, is presented below.

$$P = \{p_1, p_2, p_3, p_4, p_5, p_6\}$$

$$D = \{d_1, d_2, d_3, d_4, d_5, d_6\}$$

$$T = \{t_1, t_2, t_3\}$$

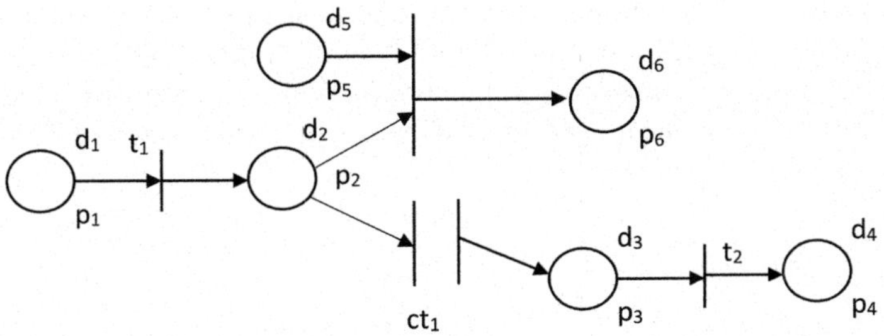

Fig. 7.1 EPN representing the knowledge base

$$CT = \{ct_1\}$$

$$I(t_1) = \{p_1\}$$

$$I(ct_1) = \{p_2\}$$

$$I(t_2) = \{p_3\}$$

$$I(t_3) = \{p_2, p_5\}$$

$$O(t_1) = \{p_2\}$$

$$I(ct_1) = \{p_3\}$$

$$O(t_2) = \{p_4\}$$

$$O(t_3) = \{p_6\}$$

$$n(p_1) = n_1$$

Say, '1' represents the truth value of variable d_1 and similarly for other variables.

Definition 2. Vector *N(t)* is called the *truth vector*, its *i*-th component represents the truth value of variable d_i, mapped at place p_i, at time *t*. The number of components of vector *N(t)* is equal to the number of places in the EPN.

After formation of the network, to start the reasoning process [5, 6], the network is initialised by assigning truth value only to the axioms (having no input arc). The places which are to be initialised are purely domain-dependent. In the reasoning process, *tt* and *ctt* are computed synchronously at the transitions and the negate transitions, respectively following which the truth values at the places are updated, which is synchronous too. The whole process together constitutes a forward computation cycle (FCC). The process is continued until the truth value at all places is stable i.e. same at two successive forward computation periods. In a pointwise notation, the computation of *tt, ctt* at the transitions and the negate transitions are described, using Eqs. (7.1) and (7.2), respectively.

7.2.2 Truth Value Updation at Transitions

The condition of firing a rule is checked first followed by which tt and ctt are computed in the EPN. A transition is called enable provided all its input places possess valid truth value which occurs due to either proper initialization of the input place in case they are axioms (having no input arc) or from the computation process in case they are intermediate places of the EPN. An enabled transition fires generating *tt* at its output arc. With reference to Fig. 7.2, the $tt_j(t+1)$ at transition t_j is computed using Eq. (7.1).

$$tt_i(t+1) = \wedge_{i=1}^{m}(n_i(t)) \tag{7.1}$$

where $\wedge_{i=1}^{m}(n_i(t))$ represents the logical 'AND' of the truth value of m number of places, i.e. $(p_1,p_2,..,p_m) \in \mathrm{I}(t_j)$.

A negate transition is enabled provided its input place possesses a valid truth value propagated from its predecessor place. An enabled negate transition generates *ctt* at its output arc, otherwise blocks further propagation of signal, i.e. the output arc becomes open. With reference to Fig. 7.3, the $ctt_j(t+1)$ at transition ct_j is computed using Eq. (7.2).

$$\begin{aligned} \mathrm{ctt}_j(\mathrm{t}+1) &= \neg\, \mathrm{n}_k(\mathrm{t});\ \text{if } \mathrm{n}_k(\mathrm{t}) = \mathrm{n}_v(\mathrm{t}-1), \text{ where } \mathrm{p}_k \in \left\{\mathrm{I}(\mathrm{ct}_j), \mathrm{O}(\mathrm{t}_u)\right\} \text{ and } \mathrm{p}_v \in I(\mathrm{t}_u) \\ &= \infty;\ \text{otherwise} \end{aligned} \tag{7.2}$$

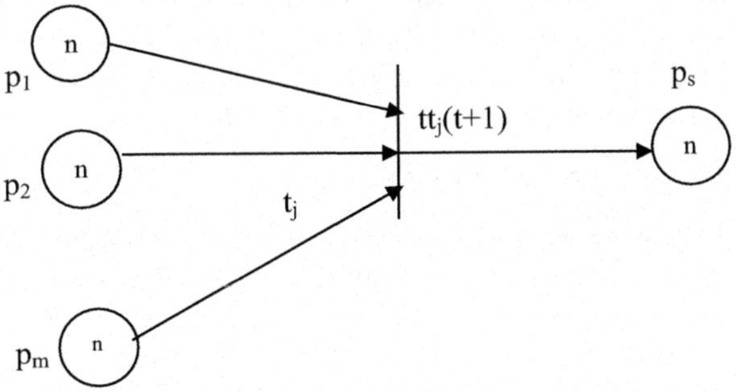

Fig. 7.2 Transition updating process

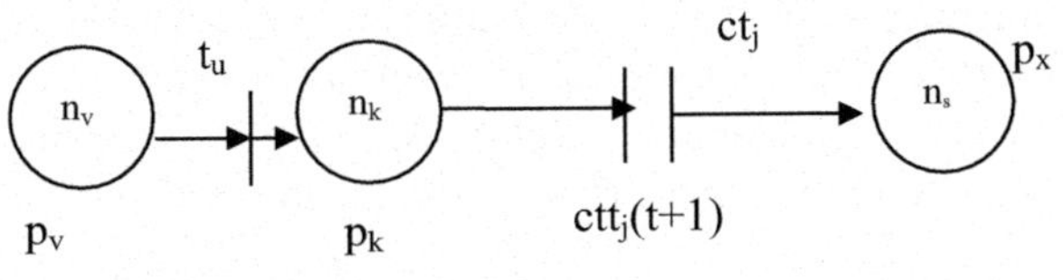

Fig. 7.3 Negate transition updating process

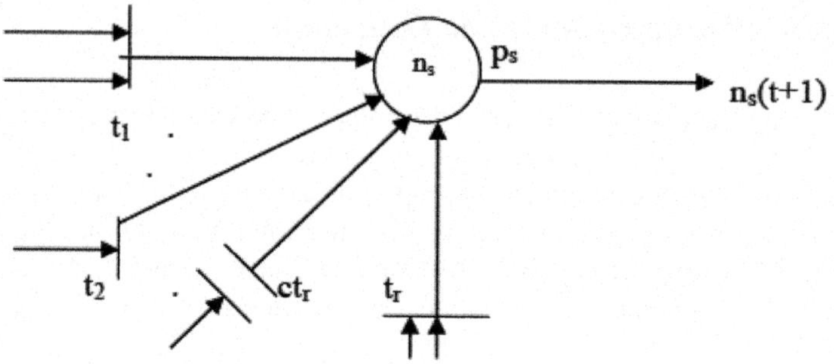

Fig. 7.4 Truth value updating process at places

7.2.3 Truth Value Updation at Places

The computation of truth value at place $p_s(n_s(t+1))$, vide Fig. 7.4, is carried out using Eq. (7.3).

$$ns(t+1) = Vtt_i(t+1)_{i=1}^{r} \vee Vctt_i(t+1)_{i=1}^{q} \tag{7.3}$$

where $ps \in O(t1, t2, \ldots, tr, t1, ct2, \ldots, ctq)$ and 'V' represents logical OR operation.

Finally, $N(\mathrm{t}+1) = [n_1, n_2, \ldots, n_z]^{\mathrm{T}}{}_{(zx1)}$, where z is the total number of places in the EPN. Thus, the current truth value of all places in the entire network are computed from their last value. The minimum value of $t = \tau$, when $N(\mathrm{t}+1) = N(\mathrm{t})$, is called the *equilibrium time* and N(τ) is termed as the steady-state *truth vector* [7]. At steady state, the truth value of the concluding places (no output arc) generates inferences, based on which the grinding parameters of the mill have been revised automatically for achieving the desired product range.

Execution Steps of EPN

Execution steps with reference to Fig. 7.1 are summarized in Table 7.1 where p_1 and p_5 are axioms and initialized with truth value, say $n_1(0) = 1$ and $n_5(0) = 1$. However, the truth value of the intermediate places i.e. p_2, p_3, p_4 and p_6 are not valid at $t = 0$.

Applying Eqs. (7.1), (7.2) and (7.3), truth value at the transitions and the places of the EPN are computed and summarized in Table 7.1.

After 3rd FCC, the EPN becomes stable and generates the inference with physical interpretation: *load the dozer* and *no need to reinitialize the motor*. The phenomenon exactly matches with the ball mill operation. It has been observed that the number of FCC for obtaining stability [8] in the EPN equals the sum of the transitions and the negate transitions residing on the longest reasoning path of the network.

Table 7.1 Steps of execution of the EPN

Initialization: $n_1(0) = 1$, $n_5(0) = 1$		
FCC (forward computation cycle)	*Truth values of enable transitions*	*Truth values at intermediate places*
1	$tt_1(1) = 1$, $tt_3(1) = 0$ -do-	– $n_2(1) = 1$, $n_3(1) = 0$, $n_4(1) = 0$, $n_6(1) = 0$
2	$tt_3(2) = 1$, $ctt_1(2) = 0$ -do-	-do- $n_2(2) = 1$, $n_3(2) = 0$, $n_4(2) = 0$, $n_6(2) = 1$
3	$tt_2(3) = 0$ -do-	-do- $n_2(3) = 1$, $n_3(3) = 0$, $n_4(3) = 0$, $n_6(3) = 1$

7.3 Case Study

To automate the grinding process, the rule base of the machine has been designed from the experimental results and implemented using EPN, described in Fig. 7.5.

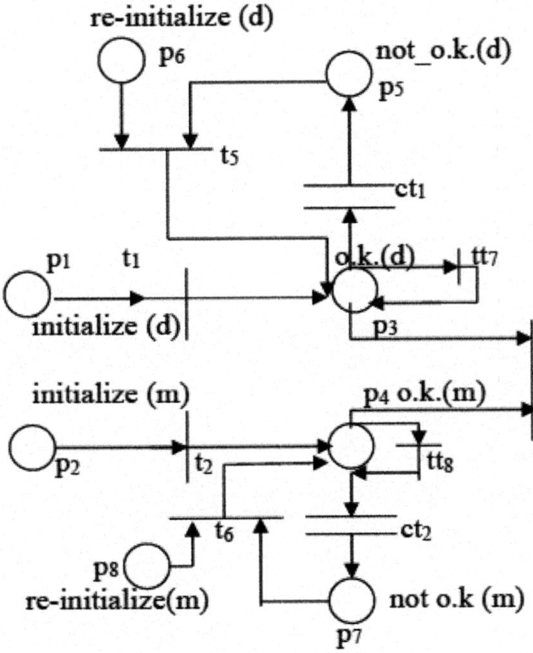

Fig. 7.5 EPN illustrating dozer section of the ball mill

The rule base [9, 10] adopted from a ball mill (B.M.) is represented in pseudo code:

```
Start:
Check (dozer); check (motor);
begin
   if (dozor is not_ok) then (reinitialization dozer) ;
      if (motor is not_ok) then (re-initialization motor) ;
        if (dozer_ok) AND (motor_ok) then
          (load_dozer);
            get (weight_ from_ load_sensors) ;
              compute (sum_weight) ;
   end;
     send (sum_weight_to_P.E.); // P.E (Processing element )
Initialization:
Initialize (motor_B.M.) ; //initialize mill motor
if (B.M._state = idle) then (B.M._status=rectify);
if (B.M._state = jammed) then (B.M._status =stop);
if (B.M._state =running) then (B.M._status = o.k);
if (B.M._status = rectify OR stop) then
 begin
   stop (motor_B.M.);
   check (motor_B.M.);
   go to Initialization;
 end;
if (B.M._status = o.k.) then (Ball-mill-working) ;
co-begin
 get (output);
 get (fr); // fr=frequency-range
   if (fr≥1800) then (B.M._status.= empty);
   if (1500<fr<1800) then (B.M._status= unknown);
   if (200<fr<300) then (B.M._status.= flotation cell);
   if (B.M._status= empty OR unknown OR flotation cell) then
begin
   stop (mill);
   check (mill);
   go to Initialization
 end;
if (600<fr≤1500) then
 begin
```

```
    (B.M._status=running with load varying);
    Ball-mill-working;
  end;
 co-end;
end.
```

7.4 The EPN Model for Dozer Section

Dozers are engaged to feed materials into the ball mill while motors and load sensors are used corresponding to each dozer to prepare the input mix of different size ranges of particles with different amounts. There may be defects in the dozers itself, as well as in the motors. Loading of materials to the mill is possible, only when the dozers and respective motors work properly. Different actions of the dozers are specified below, where '*d*' stands for dozer.

Initialise (*d*): Implies initialisation of dozer.

o.k.(*d*): Implies the dozer has no problems for dozing.

Re-initialise (*d*): Implies that dozer has some problem, so re-initialisation is necessary for dozing.

Action of the motors are similar to that of dozers, where '*m*' stands for motor:

Initialise (*m*), *o.k.*(*m*) and *reinitialise* (*m*) have the same meaning as dozer.

Load (*d*): Represents loading of dozer and is possible if dozers and the respective motors are in working condition.

On successful dozing, weight of each size range of particles are obtained from the load sensors and after obtaining individual weight, the total weight, i.e. sum_weight is computed by *P* and finally the sum_weight is fed to the ball mill. The following rule base of the dozer section is therefore, represented by the EPN of Fig. 7.5.

```
check (dozer); check (motor);
begin
  if (dozor is not_o.k) then (re-initialization dozer);
    if (motor is not_o.k) then (re-initialization motor) ;
      if (dozer_o.k.) AND (motor_o.k.) then
        load (dozer);
  end;
```

In Fig. 7.5, self-loops are incorporated around p_3 and p_4 so that any other source could not alter the valid truth values. Once the dozers and the motors start to work perfectly, self-loops ensure their proper functioning till the input signals remain unchanged.

Table 7.2 Initial values for truth value updation process

n_1	n_2	n_6	n_8
1	1	0	0
0	0	1	1
1	1	1	1
1	0	1	0
0	1	0	1

The EPN depicted in Fig. 7.5 has four axioms, namely p_1, p_2, p_6 and p_8. If the dozer and the motor are initialised properly then the truth values of p_1 and p_2 are set to one at the time of initialisation, i.e. $n_1(t = 0) = 1$ and $n_2(t = 0) = 1$, otherwise, zero. Since there are four axioms, sixteen possible input signals can be used for exhaustive testing purposes. But from the principle of truth value updation process some combinations may be treated as redundant. By considering only unique combinations, the set of initial values considered for the truth value updation process is given in Table 7.2.

Case I

Table 7.3 describes the execution steps with reference to Fig. 7.5, using Eqs. (7.1), (7.2) and (7.3). The process of updation continues till $N[t]$ becomes equal to $N[t + 1]$ for any positive integer of t.

After 3rd FCC, it has been observed that the truth values of $N[2]$ and $N[3]$ are same, therefore, at $t = 2$, the system becomes stable and $N[2] = [1, 1, 1, 1, 0, 0, 0, 0, 1]^T$ is the steady state truth vector.

Thus, truth values of p_i where $i = 1, 2, \ldots 9$ at stable state exactly matches with the functionality of the dozing section of the mill with the physical interpretation: *If the dozer*(p_1) *works properly* (p_3) *and the motor*(p_2) *of the dozer is also in working condition* (p_4) *then loading* (p_9) *of dozer is possible.*

At $t = 4$, the system becomes stable and N[4] gives the steady state truth vector.

Thus, values of p_i ($i = 1,2,\ldots 9$) at stable state become $[0, 0, 1, 1, 0, 1, 0, 1, 1]^T$ respectively, the physical interpretation can be described as:

Table 7.3 Steps of execution of the EPN considering Fig. 7.5 for Case I

Initialization: $n_1 = 1, n_2 = 1, n_6 = 0, n_8 = 0$		
FCC	*Truth values of enabled transition*	*Truth values at intermediate places*
1	$tt_1 = 1, tt_2 = 1$ -do-	– $n_3 = 1, n_4 = 1$
2	$tt_1 = 1, tt_2 = 1, ctt_1 = 0, ctt_2 = 0, tt_3 = 1, tt_7 = 1, tt_8 = 1$ -do-	-do- $n_3 = 1, n_4 = 1, n_5 = 0, n_7 = 0, n_9 = 1$
3	$tt_1 = 1, tt_2 = 1, ctt_1 = 0, ctt_2 = 0, tt_3 = 1, tt_7 = 1, tt_8 = 1$ -do-	-do- $n_3 = 1, n_4 = 1, n_5 = 0, n_7 = 0, n_9 = 1$

If initially the dozer (p_1) *has some problems and the motor* (p_2) *has some problems Then successful re-initialisation of the dozer* (p_6) *and the motor* (p_8) *results in perfect loading of dozers* (p_9).

Case II

Truth vector at $t = 3$, i.e. $[1, 1, 1, 1, 0, 1, 0, 1, 1]^T$ same as truth value of p_i s at $t = 2$ and therefore, results stability in the system for initialization: $n_1 = 1$, $n_2 = 1$, $n_6 = 1$ and $n_8 = 1$ i.e. which perfectly matches with the real situation:

Simultaneous initialization and re-initialization processes have the same effect that of Case II *and Case* III.

Table 7.4 describes the execution steps with reference to Fig. 7.5.

Case III

The system attains stable state at $t = 3$ with $n_9 = 0$, so loading of dozer is never possible in this combination, which matches the physical behaviour of the mill:

loading of the dozer is not possible when either the dozer is not in good working condition or due to improper initialization and re-initialization process.

So, after studying all possible cases it has been observed that the loading of dozer is possible and become stable only when $n_1 = 1$, $n_2 = 1$, $n_6 = 0$, $n_8 = 0$ or $n_1 = 0$, $n_2 = 0$, $n_6 = 1$ and $n_8 = 1$, which matches the physical phenomena of dozing.

Table 7.5 describes the execution steps with reference to Fig. 7.5.

Table 7.4 Steps of execution of the EPN considering Fig. 7.5 for Case II

Initialization: $n_1 = 0$, $n_2 = 0$, $n_6 = 1$, $n_8 = 1$		
FCC	Truth values of enabled	Truth values at intermediate transitions places
1	$tt_1 = 0$, $tt_2 = 0$ -do-	– $n_3 = 0$, $n_4 = 0$
2	$tt_1 = 0$, $tt_2 = 0$, $ctt_1 = 1$, $ctt_2 = 1$, $tt_3 = 0$, $tt_7 = 0$, $tt_8 = 0$ -do-	-do- $n_3 = 0$, $n_4 = 0$, $n_5 = 1$, $n_7 = 1$, $n_9 = 0$
3	$tt_1 = 0$, $tt_2 = 0$, $ctt_1 = 1$, $ctt_2 = 1$, $tt_3 = 0$, $tt_7 = 0$, $tt_8 = 0$, $tt_5 = 1$, $tt_6 = 1$ -do-	-do- $n_3 = 1$, $n_4 = 1$, $n_5 = 1$, $n_7 = 1$, $n_9 = 0$
4	$tt_1 = 0$, $tt_2 = 0$, $ctt_1 = 0$, $ctt_2 = 0$, $tt_3 = 1$, $tt_7 = 1$, $tt_8 = 1$, $tt_5 = 0$, $tt_6 = 0$ -do-	-do- $n_3 = 1$, $n_4 = 1$, $n_5 = 0$, $n_7 = 0$, $n_9 = 1$
5	$tt_1 = 0$, $tt_2 = 0$, $ctt_1 = 0$, $ctt_2 = 0$, $tt_3 = 1$, $tt_7 = 1$, $tt_8 = 1$, $tt_5 = 0$, $tt_6 = 0$ -do-	-do- $n_3 = 1$, $n_4 = 1$, $n_5 = 0$, $n_7 = 0$, $n_9 = 1$

Table 7.5 Steps of execution of the EPN, considering Fig. 7.5 for Case III

Initialization: $n_1 = 1, n_2 = 1, n_6 = 1, n_8 = 1$		
FCC	Truth values of enabled transitions	Truth values at intermediate places
1	$tt_1 = 1, tt_2 = 1$ -do-	-do- $n_3 = 1, n_4 = 1$
2	$tt_1 = 1, tt_2 = 1, ctt_1 = 0, ctt_2 = 0,$ $tt_3 = 1, tt_7 = 1, tt_8 = 1, tt_5 = 0, tt_6 = 0$ -do-	-do- $n_3 = 1, n_4 = 1, n_5 = 0, n_7 = 0, n_9 = 1$
3	$tt_1 = 1, tt_2 = 1, ctt_1 = 0, ctt_2 = 0,$ $tt_3 = 1, tt_7 = 1, tt_8 = 1, tt_5 = 0, tt_6 = 0$ -do-	– $n_3 = 1, n_4 = 1, n_5 = 0, n_7 = 0, n_9 = 1$

Case IV

Table 7.6 describes the execution steps with reference to Fig. 7.5.

The system attains stable state at $t = 3$ with the interpretation: loading is not possible when either motor is not in good working condition or due to improper initialization and re-initialization process.

Case V

Table 7.7 describes the execution steps with reference to Fig. 7.5.

By applying the truth value updation process, one can determine valid truth values at places for which stable functioning of the mill is ensured. The same can be applied for the entire ball mill, but as there are many numbers of places and transitions the whole process if traced even by tabular method, would be very cumbersome, so a simulation program is written to incorporate the truth value updation process for the entire ball mill. In the following section, the EPN model for the entire ball mill and the analysis of functioning on simulation results is described.

Table 7.6 Steps of execution of the EPN, considering Fig. 7.5 for Case IV

Initialization: $n_1 = 1, n_2 = 0, n_6 = 1, n_8 = 0$		
FCC	Truth values of enabled transitions	Truth values at intermediate places
1	$tt_1 = 1, tt_2 = 0$ -do-	– $n_3 = 1, n_4 = 0$
2	$tt_1 = 1, tt_2 = 0, ctt_1 = 0, ctt_2 = 1,$ $tt_3 = 0, tt_7 = 1, tt_8 = 0$ -do-	-do- $n_3 = 1, n_4 = 0, n_5 = 0, n_7 = 1, n_9 = 0$
3	$tt_1 = 1, tt_2 = 0, ctt_1 = 0, ctt_2 = 1,$ $tt_3 = 0, tt_7 = 1, tt_8 = 0, tt_5 = 0, tt_6 = 0$ -do-	-do- $n_3 = 1, n_4 = 0, n_5 = 0, n_7 = 1, n_9 = 0$
4	$tt_1 = 1, tt_2 = 0, ctt_1 = 0, ctt_2 = 1,$ $tt_3 = 0, tt_7 = 1, tt_8 = 0\ tt_5 = 0, tt_6 = 0$ -do-	-do- $n_3 = 1, n_4 = 0, n_5 = 0, n_7 = 1, n_9 = 0$

Table 7.7 Steps of execution of the EPN considering Fig. 7.5 for Case V

Initialization: $n_1 = 0$, $n_2 = 1$, $n_6 = 0$, $n_8 = 1$		
FCC	Truth values of enabled transitions	Truth values at intermediate places
1	$tt_1 = 0$, $tt_2 = 1$ -do-	– $n_3 = 0$, $n_4 = 1$
2	$tt_1 = 0$, $tt_2 = 1$, $ctt_1 = 1$, $ctt_2 = 0$, $tt_3 = 0$, $tt_7 = 0$, $tt_8 = 1$ -do-	-do- $n_3 = 0$, $n_4 = 1$, $n_5 = 1$, $n_7 = 0$, $n_9 = 0$
3	$tt_1 = 0$, $tt_2 = 1$, $ctt_1 = 1$, $ctt_2 = 0$, $tt_3 = 0$, $tt_7 = 0$, $tt_8 = 1$, $tt_5 = 0$, $tt_6 = 1$ -do-	-do- $n_3 = 0$, $n_4 = 1$, $n_5 = 1$, $n_7 = 0$, $n_9 = 0$
4	$tt_1 = 0$, $tt_2 = 1$, $ctt_1 = 1$, $ctt_2 = 0$, $tt_3 = 0$, $tt_7 = 0$, $tt_8 = 1$, $tt_5 = 0$, $tt_6 = 1$ -do-	-do- $n_3 = 0$, $n_4 = 1$, $n_5 = 1$, $n_7 = 0$, $n_9 = 0$

Figure 7.6 represents the EPN model of the entire ball mill, where 27 places (p_i) [$i = 1$ to 27], 28 transitions (t_j and tt_j) and 4 negate transitions (ct_j) [$j = 1$ to 4] are used to describe the functioning of the mill. For simulation purposes two different input signals i.e. '1' and '0' are considered at the axioms and Eqs. (7.1), (7.2) and (7.3) are used to propagate the signals through internal nodes, i.e. from transition to place and place to transition. To obtain the stable solution using different input signals, a simulation program has been written. Some checkpoints are defined within the network to test the functioning of the system. For instance, input signals (truth value at place p_i is denoted by n_i, for $i = 1$ to 27) corresponding to $n_1 = 1$, $n_3 = 1$, $n_{12} = 1$ and $n_{24} = 1$, the output (n_{27}) becomes 1, which accurately describe the functioning of the mill:

if the dozer is in working state ($n_1 = 1$, $n_3 = 1$), if B.M. is initialized properly then while the mill emits sound in between 600 *and* 1500 Hz, *desired output is generated from the mill.*

7.5 Summary

The hybrid system developed in the work exploiting the synergism between reasoning and learning and provides closed loop grinding in open loop condition by utilising the effect of particle size distributions. Applying reasoning in the expert control system, inference has been generated which subsequently has been converted into control actions to adapt the grinding parameters of the mill using supervised learning technique. The expert control system is not only valid for ball mills but also expected to work well with other different kinds of reactors/crushers/compound machines. The Chapter demonstrates how Petri net has been used successfully for implementing, monitoring and controlling the activity of a tumbling mill to achieve optimum grinding within residence time.

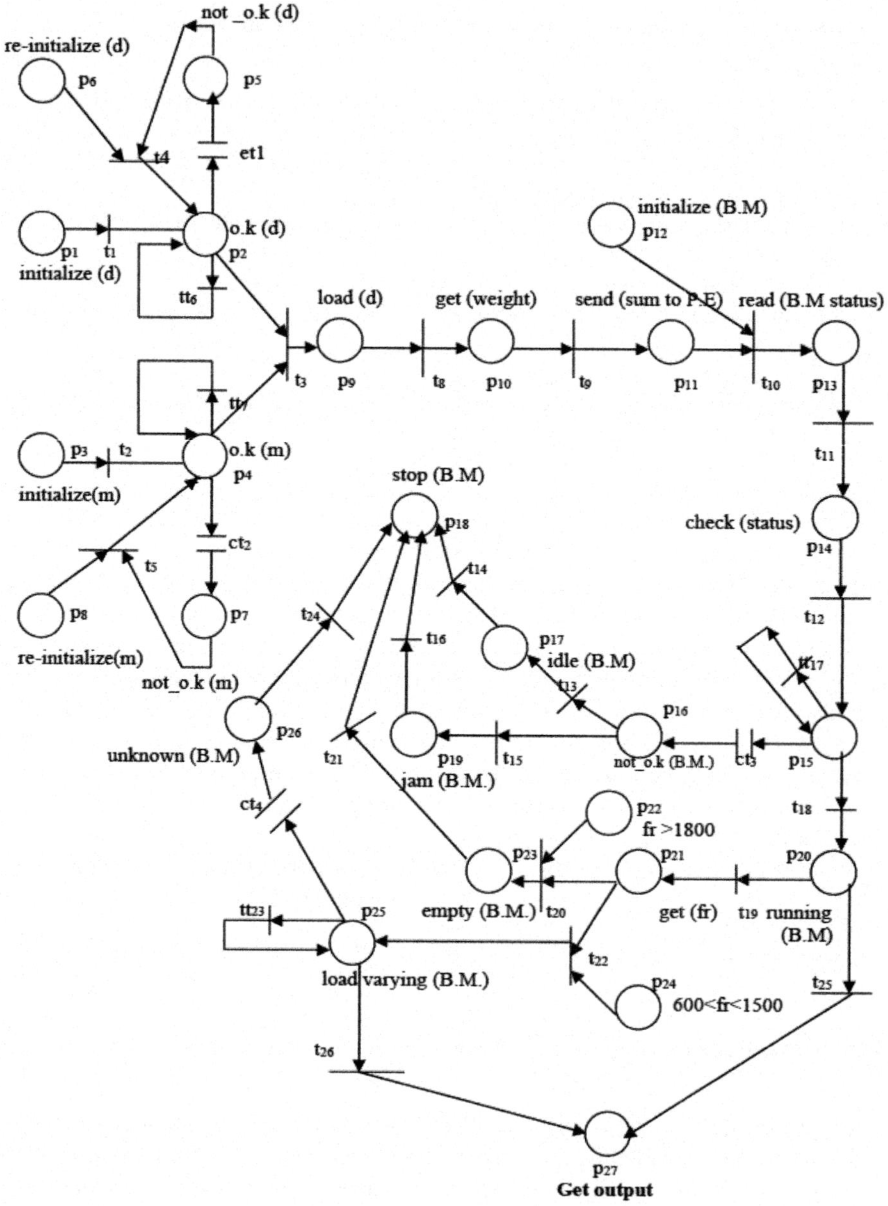

Fig. 7.6 EPN model for the entire ball mill

Exercise

1. Differentiate the functioning of Extended Petri Net from a conventional Petri Net?
2. What is the difference between a Petri Net and a finite state machine (FSM)?
3. Describe Truth Value updating process.
4. What are the steps to execute an Extended Petri Net?
5. How can Petri Nets be used for workflow modeling and Stability also?

References

1. Lindemann, C.: Performance modelling with deterministic and stochastic Petri nets. With CD-ROM (1998). https://doi.org/10.1145/288197.581195
2. Wang, J.: Petri nets for dynamic event-driven system modeling. In: Handbook of Dynamic System Modeling (2007)
3. Peterson, J.L.: Petri net theory and modeling of systems. Prentice Inc., Englewood Cliffs, NJ (1981)
4. Bhaumik, A., Sil, J., Banerjee, S.: Designing of Intelligent Hybrid Control System for Tumbling Mill Operation. CIT (2003)
5. Sil, J., Konar, A.: A hybrid approach to knowledge acquisition using Neural Petri nets and DS theory. IJCIA **1**(2), 203–223 (2001)
6. Sil, J., Konar, A.: Reasoning using a probabilistic predicate transition net model. Int. J. Model. Simul. **21**(2), 155–168 (2001)
7. Kasturia, E., Dicesare, F., Desrochers, A.A.: Real-time control of multilevel of manufacturing systems using color Petri nets. In: Proceedings of IEEE International Conference Robotics and Automation, PA, pp. 1114–1119 (1988)
8. Genrich, H., Lautenbach, K.: System modeling with high level Petri nets. Theoret. Comput. Sci. **13**, 109–136 (1981)
9. Looney, C.G.: Fuzzy Petri nets for rule-based decision-making. IEEE Trans. Syst. Man Cybern. **18**(1), 178–183 (1988)
10. Waterman Donald, A.: A guide to expert systems. Addision Wesley (1986)

Chapter 8
Conclusion and Future Scope

8.1 Discussions

This chapter concludes the book and provides the future research path. This volume aims to introduce a new paradigm towards controlling the grinding procedure of a ball mill using computational methods. The researcher gets a new dimension to use Acoustic Signature to predict the states of grinding, including the End state when the ball mill automatically comes to a standstill condition and users achieve the desired particle size distribution. This book also offers a wide spectrum of research work done to design an Intelligent system with Learning and Mathematical modelling.

The development pathway is discussed in this book with a goal to design an intelligent control system for automatic operation of the ball mill and its application in the real world scenario. The encouraging results obtained from different frameworks presented in the volume highlighting the importance of the proposed methods in the mining industry. The fundamental work is to predict the working of the mill and simulate the on-chip system to control the mill at runtime. In this Chapter, the revisit of different findings against the aim and challenges towards the designing of an intelligent system to control the ball mill operation is discussed. Finally, this Chapter discusses the directions of future research in this field.

Here the focus of each Chapter is discussed. In the Chapters, the simulated results based on different computational models are verified with the experimental outcomes, which make the models strong enough to be deployed for real time applications.

- Chapter 1 introduces to the reader the domain of research work which mainly focuses on the mining industry which has ample scope to automation. In Chap. 1, discussion is on the mining operation, how at present such operations are handled manually by using ball mills. The problems of the existing systems are also highlighted which motivates the researchers with an aim to minimize the wastage of materials. The designing preliminaries are discussed in this volume elaborately,

J. Sil et al., *Intelligent Real-Time Control Systems to Predict the Particle Size of Minerals Using Acoustics of Ball Mill*, Studies in Systems, Decision and Control 600,
https://doi.org/10.1007/978-3-031-91290-0_8

for the better understanding to the readers that includes students, researchers and real time application in the mining industry.

- Chapter 2 is dedicated to highlight how the acoustics of the Ball Mill are utilized to develop the controller. During the grinding operation how the sensors are installed and acoustic signals are captured are described. During the grinding operation, the acoustics of the running ball mill loaded with different raw materials with particle size distribution and empty conditions are captured to prepare a database and this can be used for later analysis. Finally, a method is discussed to differentiate the running ball mill in its empty and loaded condition.
- Chapter 3 mainly focuses on particle breakage modelling. A '*K*' factor is determined based on which a relationship has been derived between the breakage characteristics of different kinds of materials. Experimental analysis of the data reveals some interesting results that help to predict the particle size distribution of materials obtained from grinding. A new supervised training algorithm is also used in this Chapter to obtain input (feed) distribution of materials for a desired output product distribution. Experimental data are verified with simulation results to establish the predictive grinding using the proposed algorithm.
- Chapter 4 is dedicated to propose two efficient strategies to predict different states of a running ball mill in different load varying conditions. The first method is established based on a mathematical model to obtain the acoustic patterns at different states of transitions of the running ball mill by analyzing the sound of the ball mill in its load varying conditions. The second method predicts the terminating condition of the ball mill by extracting the statistical features using time domain information of the acoustics.
- Chapter 5 discusses how to determine the residence time of breakage of the materials in the ball mill in order to forecast any state of grinding of the ball mill and correlate the same with the particle size distribution obtained experimentally. Using Genetic Algorithm the desired state of grinding of the ball mill is predicted based on its two previous consecutive states and after running the ball mill up to that predicted state, it is stopped automatically. The particle size distribution of the materials at different states of grinding are obtained by applying Gaussian Mixture Model (GMM) and clustered using fuzzy C-means (FCM) clustering algorithm where number of groups are equal to the predetermined states of the ball mill. When the mill stops, the *psd* of the breakage material corresponding to a particular state must belong to the respective cluster, verifying the experimental result.
- Chapter 6 discusses a research background to design a real-time Intelligent system to predict the states of crushing, including the End State of the loaded running ball mill. At the end of crushing, the mill stops its crushing operation, and the user obtains expected particle size distribution (*psd*). This Chapter proposes an intelligent system, which in real-time with some predefined control parameters achieves desired *psd* of a specific input mix, and stops the mill automatically. The system on chip can be designed so that the machine can be stopped automatically whenever the desired particle size distribution has been achieved. The proposed

approach is based on the analysis of the time series data and validated using Machine learning technique.

- Chapter 7 demonstrates how Petri Net has been successfully applied for implementing, monitoring, and controlling the activity of a ball mill to achieve optimum grinding within residence time. To simulate the entire scheme of a ball mill, a knowledge base in terms of If–Then rules are framed describing the grinding operation of the ball mill. The rules are realized by an extended-Petri-net (EPN) structure to generate the operational status of the ball mill using the reasoning process. The inferences generated from the reasoning process are converted into control actions to achieve the desired product. The expert control system is not only valid for ball mills but is also expected to work well with other different kinds of reactors/crushers/compound machines.

8.2 Future Scope

In this book the methods discussed for the ball mill crushing operations, are based on the ball mill acoustics. Synergisms of different Soft Computing approaches and Mathematical modelling are developed using the parametric approaches and it has been proved that this follows the experimental results. Different traditional machine learning and deep learning models are beneficial to handle large datasets and extract deep knowledge, providing excellent performance. The aspect of runtime of the system can be explored in future by designing an on chip system to control the operation of the ball mill. The methods discussed in this volume are for a specific type of mill operation, but in future the automation of the other types of mills can also be analysed.

The manufacturer's authorised representative in the EU is Springer Nature Customer Service Centre GmbH, Europaplatz 3, 69115 Heidelberg, Germany. If you have any concerns regarding our products, please contact ProductSafety@springernature.com

Printed and bound by CPI Group (UK) Ltd, Croydon, CR0 4YY
30/09/2026
02225604-0015